Benjamin Smith Lyman

Geological Survey of Hokkaido

Report of Progress of the Yesso Geological Surveys for 1875 and Seven Coal Survey

Reports

Benjamin Smith Lyman

Geological Survey of Hokkaido
Report of Progress of the Yesso Geological Surveys for 1875 and Seven Coal Survey Reports

ISBN/EAN: 9783337060459

Printed in Europe, USA, Canada, Australia, Japan

Cover: Foto ©berggeist007 / pixelio.de

More available books at **www.hansebooks.com**

GEOLOGICAL SURVEY OF HOKKAIDO.

REPORT OF PROGRESS

OF THE

YÉSSO GEOLOGICAL SURVEYS

FOR

1875,

AND

SEVEN COAL SURVEY REPORTS.

BY

BENJAMIN SMITH LYMAN,

CHIEF GEOLOGIST AND MINING ENGINEER.

TOKEI :
PUBLISHED BY THE KAITAKUSHI.
1877. -

CONTENTS.

VIII

TABLE OF ERRATA.

SOME OF THE MORE IMPORTANT ERRATA.

Page		line		from bottom,	for :		read :	
Page	5,	line 14	from bottom,	for :	are,	read :	we.	
,,	6,	,, 14	,, ,,	,,	to	,,	to be.	
,,	6,	,, 1	,, ,,	,,	but	,,	but it.	
,,	7,	,, 11	,, ,,	,,	cauldrous	,,	cauldrons.	
,,	18,	,, 6 ,,	,, ,,	,,	secrue	,,	secure.	
,,	29,	,, 6...	... ...,,		to	,,	at.	
,,	39,	,, 5 from bottom,	,,		case	,,	ease.	
,,	74,	,, 8...	... ...,,		diameter	,,	thickness.	
,,	91,	,, 13...	... ...,,		*Shipments*	,,	*Shipment.*	
,,	104,	,, 3 from bottom,	,,		3,500,000	,,	1,800,000.	
,,	106,	,, 8...	... ...,,		level	,,	level; and	

the shape of the coal bed by similar but broken lines one hundred feet apart in level.

,, 126, ,, 15... ... for : *shipments* read : *shipment.*

SOME OF THE MORE IMPORTANT ERRATA IN B.S.L.'s REPORTS IN THE VOLUME OF KAITAKUSHI REPORTS OF 1874.

Page		line		from bottom,	for :		read :	
Page 337,		line 6	from bottom,	for :	Sapporo,	read :	Sapporo.	
,,	373,	,, 2	,, ,,	,,	side,	,,	other side.	
,,	447,	,, 5	,, ,,	,,	with-	,,	without.	
,,	460,	,, 1	,, ,,	,,	6½	,,	86½.	
,,	464,	,, 4	,, ,,	,,	letter	,,	better.	
,,	467,	,, 17	,, ,,	,,	even	,,	eleven.	
,,	471,	,, 4	,, ,,	,,	That would	,,	At one half that it would.	
,,	480,	,, 5...	...,,		their	,,	or their.	
,,	481,	,, 8...	...,,		as	,,	as is.	
,,	598,	,, 4...	...,,		much	,,	much oil.	
,,	616,	,, 8...	...,,		photographic,,		photograph- [ing.	
,,	620,	,, 1...	...,,		*Land*	,,	the *Land.*	
,,	146,	,, 9...	...,,		150	,,	450.	
,,	150,	,, 16...	...,,		4,300	,,	43,000.	

GEOLOGICAL SURVEY OF HOKKAIDO.

REPORT

OF

DAILY PROGRESS

OF THE

YESSO GEOLOGICAL SURVEYS OF 1875,

WITH

COMMENTS ON OTHER MATTERS,

BY

BENJAMIN SMITH LYMAN,

Chief Geologist and Mining Engineer.

TOKEI:
PUBLISHED BY THE KAITAKUSHI,
1876.

It is desirable that you should advise us, without fear, not only on that matter [the geological and mining exploration of Hokkaido] but also any matter of importance that you may think it proper for us to carry out on that Island. *Extract from a letter of the Kaitakushi to B. S. Lyman, dated 30th May, 1875.*

GEOLOGICAL SURVEY OF HOKKAIDO.

REPORT OF DAILY PROGRESS IN 1875.

Hakodate ; 22 June, 1875.

His Excellency K. Kuroda,
Kaitakuchokuwan.

Sir :—I have the honor to report that having sailed from Yokohama on the afternoon of the 15th of this month in the Steamer *Taihei Maru* accompanied by my ten geological assistants, we arrived here safely on the morning of the 20th.

On the way we stopped in Sendai Bay from about noon of the 17th until the middle of the next afternoon ; and improved the opportunity of taking a hasty glimpse of the geological, topographical and other features of that neighborhood. As the place has been seldom, if ever, visited by geologists, and as our Yesso surveys give some clue to understanding it, there may perhaps be some interest to you and to the public in a brief account of the little we could see in a few hours.

We first landed at Samusawa, a village of perhaps 500 inhabitants on an island close by the entrance of the bay and near the anchorage of our steamer. Finding there would be time for a longer excursion, eight or nine of our party took a small boat and crossed the bay westerly five ri to Shiogama, a village of perhaps 4,000 inhabitants, and spent the night there, returning in the morning to the steamer, and, as it was not yet ready to sail, revisiting Samusawa.

The steamer where it was anchored lay surrounded by islands with many steep, light-gray cliffs up to perhaps a hundred feet in height. A few hundred yards further from the sea a portion of the bay large enough to anchor a fleet is wholly land-locked and is said to be fifty feet deep. The islands are various in size, from a few yards in diameter to several miles. The highest is perhaps Matsushima, within sight of which, a couple of ri distant on the west, we passed in crossing the bay, but even that seemed not more than a thousand feet high, if so much. The view from its top, however is famous as one of the three finest in Japan, and not only covers the "eight hundred and eight islands" of the bay, but is said to reach even to Fujisan. The shores of the islands are not all cliffy, but in many places leave gentle slopes for small villages of fishermen. On one island dwellings of unknown antiquity were cut in the face of steep cliffs of the soft sand rock. They are no longer inhabited, but the square topped entrances are still to be seen, and are in some cases filled now with sacks of coal. On another island the soft greenish gray sand rock has been carved into a small sitting Booddha, wrought, as I understood, by the inhabitants of a neighboring village, and does credit to their skill. In other places the rock has been cut into posts for mooring junks. There are many places where the sea has cut holes through from one side of an island to the other leaving natural bridges ; and such work is favored by the fact that the rock is in layers that are not very steeply inclined, say ten or twenty degrees only.

The rock at the very first seemed much like what we had seen in Yesso ; both in color, very light greenish gray, here and there weathered light brown or buff, and in texture, that of a soft, rather coarse sand rock. At the Samusawa landing place we noticed a block of stone that had been quarried within a ri of there, and seemed to be of the same formation as the rocks of the cliffs already mentioned ; and it contained a bit of coal, black and shining and closely resembling the coal of Yesso. It seemed therefore highly probable that the rocks of the Bay were of the same age as the coal-bearing rocks of Yesso which I have called the Horumui Group ; and we were still more assured of the fact when we found that even in the direction of the strike, north 45 or 50 degrees east there was a perfect correspondence. The dip throughout the bay was almost every where south easterly, but at one or two points a small roll in the rocks caused a dip in the opposite direction for a short space. At Shiogama are found some layers of the rock to be made up in great part of more or less angular pebbles, which proved to be of trachytic porphyry with what looked like sanidin crystals. At other points in the bay we afterwards found larger beds of such pebble rock that looked much like the tufaceous beds that overlie the coal-bearing rocks near Kayanoma (Iwanai); a still further confirmation of our views as to the age of the Sendai rocks. Moreover, just as we were finally leaving Samusawa a block of stone was seen that had imbedded in its surface a large number of small rather fresh looking oyster shells, that again called to mind what we had seen among the rocks of the Horumui group. Indeed, the identity of age could no longer be doubted and the resemblance in details was very remarkable considering the distance of some 300 miles to the nearest exposures of the group in Yesso. It is also impossible to doubt now that the Hiyamidzu (Iwaki) coal, within twenty ri or so of Sendai Bay, belongs likewise to the same rock group ; and it seems extremely probable that the group may be found to extend as much further to the west of Sendai as it does to the north-east and to include therefore possibly the western Japanese coals, which resemble the Yesso coals so much in character.

Although Shiogama is but four or five ri distant from Sendai, which our boatmen thought to be a town of 100,000 inhabitants, we found it impossible to hire two jinrikishas there were in the village to take two of us there in the evening and back in the early morning, owing it was said to a previous engagement, One of our boat men had assured us that the trip could be easily made, and that two rio would be ample compensation to the four coolies that would pull us, Some doubts however were raised as to the possibility of my going so far outside of " treaty limits " without molestation from officials. At Shiogama nevertheless the inhabitants seemed to somewhat accustomed to the presence of foreiguers, and one landlady of an hotel or teahouse even came to me in the street and by mistake accosted me as an old acquaintance and invited us to patronise her house. A public notice close by, forbidding entrance into the adjoining Shinto temple grounds on horseback or in a carriage, was written in English and French as well as Japanese.

Such a prohibition seemed in this case hardly necessary, as the entrance was by a steep flight of 234 steps. The temple seems to have been formerly a famous one and is in a grove at the top of a hill. A little lower down the hill towards the sea is a large building formerly used for the dwelling of the priests, but is now chiefly empty and

ruinous but partly used as an hotel. We found there
what was called a petrified pine, and it turned out to be a
block of flint a couple of feet or more in diameter. It
was of a yellowish color, in some parts dark like pine
wood full of turpentine, in others light, almost white,
and in parts veined somewhat as if by the year rings of a
tree. But the veins did not seem to be at all circular
like such rings, nor very continuous ; and nothing like
lengthwise fibres were seem. The layers were too obscure
to leave no doubt ; but the impression given was rather
that of merely a mass of flint that had by repeated layers
filled some cavity or vein in rocks.

The great curiosity however of the village is the remains of four iron salt kettles which give rise to the
name Shiogama. The time when they were used for
making salt was so long ago, and is so completely forgotten, that some of the inhabitants even assert it to have
been before the time of Jimmu Tenno, or say three thousand years ago. But it is hardly likely that any
traces at all of the kettles could have endured in their
exposed position for a very great length of time, and
it is even surprising that they could have lasted
until their age should be so thoroughly forgotten. They
are the lower nearly flat part of what must have been
large cauldrous ; and are about three feet in diameter, a
third of a foot deep and perhaps a tenth of a foot thick.
They stand near together in the open air upon what seem
to have been furnaces for boiling the salt water, and are
in a little enclosure in the central part of the village.
They are exposed to the rain and to those alternations of
wet and dryness that are so favorable to the formation of
rust and to the rapid disappearance of iron.

Shiogama is the nearest port to Sendai, and is said to
have been formerly a place of a thousand houses, but
now to have only seven or eight hundred. The temple

probably attracts fewer visitors and inhabitants than formerly, and there are perhaps other causes for a diminution of the population. From the sea it is approached by a long, narrow, shallow, crooked channel. The steamers at present seem to stop near where ours did and to tend rather to the building up of two small villages on the neighboring islands. Between the islands the tide in flowing and ebbing produces strong currents.

A pleasing sight both at Shiogama and at Samusawa was a tall post a year or two old announcing that there, was a primary school there, a creation of the Mombusho. It is heartily to be hoped that such schools may soon become common throughout Yesso also.

The chief occupation of all the villages seemed to be fishing, and many of the inhabitants were busy cutting up the dark red flesh of the bonito (maguro) or loading whole ones upon horses, two for a load.

At the best hotel we could find in Shiogama the charge for lodging with two meals was only 16 cents! Why should keepers of public houses in Yesso, with almost identical accommodations, as I found at Otarunai two years ago, be reluctant to entertain our official parties at about double that price?

Iwanai ; 29 June.—The day of our arrival at Hakodate (20th) was Sunday and our heavy baggage was not landed until the next day, which was the Japanese official day of rest. However by the night of the 22d the needful purchases and preparations for our journey had been made, and eight of our geological assistants, Messrs. Inagaki Kuwada, Kada, Saka, Shimada, Yamagiwa, Mayeda, and Nishiyama went aboard the steamer *Capron* to sail at daylight for Mororan on the way to Sapporo. The delay at Hakodate had also given an opportunity to do some office work, such as beginning this report to yourself, making out instructions to assistants and some other writing.

It is the common remark among both natives and foreigners in Hakodate that the business of the place is very dull and apparently diminishing year by year. Perhaps the diminution, if real, only reaches back two or three years, to the time when the building of the New Road must have given a great start to the building up of the town and to its hopes. The harbor is certainly very fine, but in other respects there would seem to be no very good ground for expecting the business of the place to increase to an indefinite extent. Its prominence as one of the ports open to foreigners is fast sinking into insignificance, from the fact that the increasing number of Japanese steamers, which can visit other ports on their way coming and going, are making it unprofitable for foreign vessels to come so far to this one point. Hakodate, moreover, near the small end of the narrow mountainous southern portion of Yesso is not the natural outlet of a very large nor very productive region. The main products, seaweed and fish, can with the improved Japanese marine be shipped direct to Nippon or to China, from many other points on the peninsula or on the west of the island without any need of transhipment by foreign bottoms at Hakodate. Whatever may be said by the disappointed foreign or native property holders at Hakodate, it is clear that a town at the natural outlet (or, by railroads, artificial outlet) of a great valley like the Ishcari, rich in fertile land, in timber and in coal is far surer of indefinite growth throughout a long future. Sapporo may perhaps not be destined to be such a business centre, for which Otarunai or possibly Mororan may be better fitted; but Sapporo will always be a pleasant, healthful, picturesque and convenient place for a seat of Government in a fine farming region and very near a large marine, mercantile, agricultural, manufacturing and mining population.

On the morning of the 23d of June the two remaining

assistants, Messrs. Misawa and Takahashi, and myself accompanied by the quartermaster, Mr. Nishimura, and my copyist, Mr. Adachi, set out for Mori on our way to the Kayanoma coal mines and to Sapporo. We arrived at Mori, 29 miles (11¾ ri), towards night ; but on the way I had the pleasure of spending several hours with Mr. Dun at the Nanai Government farm and of hearing something of the very useful work he has been doing there. As I have in a former report already spoken of the geology and topography of the day's journey, and indeed of our whole journey from Hakodate hitherto, a repetition of them may be passed over here.

The New Road, by which we travelled, had had its bridges all put in good repair since last fall except one near Mori, which alone makes the road quite impassable to waggons. Although the road is generally very smooth, there are still on the mountain deep traces of the cross ruts made by the horses in the mud of winter. On that part of the road some laborers were improving the slope of the sides of the cuttings which had by washing become too steep. In these and other places the drain at the side of the road had become at least temporarily filled up, and should be thoroughly reopened. To roads quite as much to any thing else the homely proverb well applies : " A stitch in time saves nine."

The number of passers on the road seemed to be very few indeed, and the amount of goods carried across extremely little ; perhaps owing partly to the season of the year, when it is comparatively safe for vessels to go direct to Mori and to other less sheltered points. If the road however were kept in repair, there would perhaps be ample inducement for the establishment of at least one line of daily waggons for carrying goods along the roads much more cheaply than can be done on the backs of horses. Such a business would be sure to increase, from the very fact that its cheapness would cause a greater demand for the articles so transported.

At Nanai I was more than ever convinced of the advantage of the plan I proposed last year, that the Kaitakushi should give away or very cheaply sell the most of the horses of Yesso to the inhabitants. There is great ignorance about the proper care of horses, and as long as they are government property there will naturally be comparative indifference as to their right treatment. Perhaps even on the government farms (where, however, men better fitted if possible than common Yesso farmers should be selected for managing the live stock) it might be economy to present some of the officials with good horses and forbid them to ride any of the government ones. Self interest would teach the owner of horses not to abuse them unmercifully.

On the 24th of June, we rode from Mori to Yamukushinai 16¼ miles (6½ *ri*). In passing Washinoki the two geological assistants and I left our horses for a while and walked into the woods under the guidance of five bright little boys (no older guide could be obtained that day) to see the pits that were dug there last year for oil by a certain Japanese company of merchants. There was a rumor last fall, as mentioned in my recent report on the Washinoki Oil Lands, that they had succeeded in obtaining as much as 50 gallons (or 12 *to*) of oil a day. We could find no one the other day who knew the exact amount obtained, but it was very clear from what we did learn that the rumor was a gross exaggeration ; and that if any thing like that amount of oil was gathered in one day it was not the accumulation of a single day in the wells, but of several. Indeed, the most definite and probable estimate that we obtained by questioning a number of villagers was that in the whole season about 25 (more than 20 but less than 30) tubfuls of " two *to* " each, or about 200 gallons (or 50 *to*) had been obtained in the whole of last year. Six or seven pits had been dug ; all but one or two were vertical wells two or three feet square; and the exceptions were small horizontal drifts draining such shafts ; for these were all dug at some little height on the bank of the brook and generally reached down to about its level, below which the water was probably troublesome. The deepest and almost solely productive of the new wells is about 45 or 55 feet deep and at the top for twenty-five feet down about three feet square, but for the rest of the depth only about two feet square ; and like the other wells is dug apparently throughout in soft greenish gray sand rock. The yield of the deep well is said to have been two tubsful (16 gallons or 4 *to*) daily. There were traces of oil at another well, a drift running to a shaft about 30 feet deep and two feet square. The drift probably starts at a point where one of the old oil exposures was. Another similar shaft about 20 feet deep a few yards distant was quite unproductive. Another likewise without oil was dug perhaps 25 feet deep on the bank a few yards from the exposure that was described in my report on these oil lands as the second best of all. An unsuccessful shaft and drift were also dug, they say, near the main well of that report.

We have the position of the wells precisely enough to lay them down pretty exactly on our lately finished map of the oil lands, although we have not yet had a chance to do so. The new productive well will no doubt be found to be on, or very close to, the outcrop of one of the oil bearing beds as laid down on the map. We were told that the oil taken was refined for burning in lamps, and we saw a lampful of it. The price was said to have been ten cents for one quarter of a *sho* (or at the rate of 40 cents a *sho*, or a dollar a gallon.) It was sold in the neighboring villages, especially at Mori. The oil

however is so heavy that the yield of lighting oil on re-
fining must be comparatively poor ; and the best plan
would be rather to make lubricating oil. The works
were carried on from the third to the tenth month of last
year by half a dozen officials and nearly as many laborers,
but were at length finally abandoned as wholly unsuccessful, a result quite in accordance with the advice that I gave
at the outset.

At Yamukushinai we revisited the oil exposures there.
In addition to what we had seen before, the priest of the
Buddhist temple showed us near the back of the building
two places where slight traces of oil had been found in a
small ditch, and another such place at the mouth of the
ravine a few yards to the east, places that all correspond
very well with the outcrop of the oil-bearing bed as laid
down on our recent map. The small quantity found also
confirms what is said on that point in my late report on
those oil lands. Near the mouth of the ravine, on its
eastern bank, we found a small exposure of soft greenish
gray sandrock without any clear dip, and without any
oil ; although some small black spots in it may possibly
be traces of asphalt.

On the 25th of June, we rode from Yamukushinai to
Oshamambe, 22½ miles (9 ri) ; stopping a little to look
once more at the curious rock of Kuroiwa. The beaches
on either side of that village had a large number of men
upon them hauling in large nets of herring, Looking up
Kunnui valley numerous patches of snow were still to be
seen on the mountains, probably on the westerly side of
the Toshibets.

Through the morning of the 26th June, I was very ill
and it was at first a little rainy, so that travelling was
given up. But at noon the weather had cleared up, my
health was better, and a few horses were obtained, so that
a portion of our party could go on from Oshamambe

to Kuromatsunai, 15 miles (6 ri), where we arrived some time before nightfall. The road, though still very rough and bad, had by the dryness of the season without any special repairs, so much improved that a passer by could form no conception from it of its horrible condition last fall, a condition that must, however, return every year until some effective repairs are made. The travel even by so poor a road across the peninsula here seems to be far greater than between Hakodate and Mori, judging by the very large number of men and horses we met that day and the next ; and the importance of the road would certainly seem to claim for it more attention than it has for a long time received. Perhaps with a better road several waggon loads daily by some regular line could be carried across even at the beginning. The saving of distance over going round the peninsula by water is here far greater than at Mori, and the natural facilities for road building much better.

Iwanai ; 30 June.—At Kuromatsunai we met Mr. Carrey, a French traveller, who was on his way back to Hakodate from a trip to Sapporo. It seemed almost like meeting a fellow countryman so far from home, and the evening and early morning passed very agreeably in his society.

He had heard (what I have long believed to be the case) that foreigners were desirous of undertaking coal mining in Yesso, and foreigners too who would prefer to associate Japanese capital with their own. It is quite the contrary of what was stated to me lately in Yedo by one of your representatives as an argument for the government's working of the mines ; namely, that neither natives nor foreigners were willing to enter upon such undertakings, and it was necessary for the government to set an example in that line to its countrymen. It was promised, however, to refer to you my letter protesting against such govern-

ment works as sure to be wastefully or injudiciously
managed by officials, as compared with the carefulness
of men who are laying out their own money.

You charged me a year and a half ago not to divulge
the rules by which the Kaitakushi would dispose of
mining privileges to private individuals ; and it has been
easy to comply with the request, since I have been kept
to this moment in absolute ignorance of what those rules
may be, and even of any motive for not making them public. The world, however, if not the general government
nor the Kaitakushi, must hold it almost incredible that a
man in my position should not be consulted on a matter
in which I should naturally be thought to be in some
degree responsible.

I shall therefore without waiting any longer for a
special invitation venture to give here a rough general
outline of what seems to me the best manner of disposing
of coal mining rights in Yesso, aided by recollections of
the condition of coal mines in other countries. In the
United States, for example, the coal lands have been sold
with the coal in fee simple to private individuals, who
have made whatever arrangements they pleased in regard
to mining, subject only to the laws in behalf of the safety
of miners, and subject (like the holders of other property)
to government taxes. In Nova Scotia, on the other hand,
although the surface of coal lands is sold like other lands,
the right to mine the coal is reserved by the government,
but is leased to private parties at a royalty of ten cents a
ton. In case of a sale of the coal mining privileges out-
right, as has been done in the United States, the individuals
who buy would have to bear all the risks, uncertainties
and delays of mining and selling the coal, and could
hardly pay in advance an adequate price for the unmined
coal; it would necessarily be set very low, as it was there.
The government can much better afford to undertake such

risks and uncertainties and endure the delays, and in fact would have much more control over them. The government, too, in merely leasing the mining privilege would retain still more distinctly the right to inspect and regulate the mode of working, so that not only the miner's lives should be safe, but the coal should not be mined in such a way as to cause (as sometimes happens) the wanton waste of large portions of the coal beds. I am not informed whether or not the Kaitakushi has the right to alienate the coal for ever from the government ; but, even if it have, it seems most advisable to adopt the system of leasing, under proper rules, of course.

Those rules should, if possible, at the same time that they secure to the government (that is, the public) a good final return for its property, encourage mining enterprise duly among the public at large, and not to the exclusive benefit of the personal or political friends of government officials. In Nova Scotia, the coal mining privileges are leased on very liberal terms, in order to increase the business and prosperity of the country, looking not merely for direct revenue from the coal mined, but to the greater general wealth of the population. Any individual, native or foreign, can obtain the exclusive " right to search " for coal (with a view to future mining) over any tract of five square miles. not already similarly occupied, on payment of a small fee, if I remember rightly only twenty dollars. At the end of a year he may select any one of the five square miles, and on payment of a fee of fifty dollars obtain the "right to work" coal. If within two years thereafter he shall have really begun effective mining operations, he may obtain a "lease" of the coal of that square mile at the rate of ten cents a ton for every ton mined and sold. All leases are (for special local reasons) to terminate in the year 1886, when the government will once more be free to dispose of all its mines. At the

time when the present system was adopted the leases had
a little over twenty-five years to run ; and twenty years is
a common length of such leases from private owners in
the United States. There, however, the royalty is commonly much higher, twenty-five or thirty-seven and a
half cents a ton for anthracite. In Nova Scotia the
holder of a right of search is required to report the results
of his investigations to the government, and the holder of
a lease is required to hand in yearly a map of the underground workings.

The Nova Scotia system is in some of its details not
altogether without shortcomings. Many tracts after the
leases are once granted lie almost or quite unworked for
years, without adding, as they should, to the revenue of
the government, nor to the business activity of the country. Although the rights to search and to work are
nominally given to the first applicant, cases might well
arise where the friends of not over honest government
officials might by secret information, or otherwise, obtain
an advantage over others in trying to obtain such privileges. It has also been a common thing for the lucky
owner of such a privilege to sell it for a large sum of
money, especially after a successful but inexpensive search
for coal. A gambling spirit of speculation is in that way
fostered, and the inhabitants hasten to take up mining
rights in the surrounding country, although wholly
ignorant of the probable mineral value of the land. The
owners of mining rights are as negligent as they dare be
under a liberal, lenient government in regard to reporting
the results of their search or mapping their underground
works.

·It seems, therefore, decidedly best for the government
to keep in its own hands the search for minerals and the
underground mapping, which by a general geological
survey and body of surveyors and mine inspectors would

in the end cost the public much less than the aggregate
of numerous disconnected surveys of the separate tracts.
Wild speculation in lands of imaginary mineral value
would in that way be prevented with its great attendant
losses ; and the government could at once furnish intend-
ing operators with a geological and topographical map
and report of the tract to be worked. In return it would
be fair enough to demand a somewhat larger fee, say, five
hundred dollars for a square mile ; for a payment of which
fee, on the other hand, the preparation of a similar map
and report within a reasonable time might be offered for
any desired tract of like size. If effective mining operations should be actually begun within say a year (or
two years) after the transfer of the map and report, then
a lease might be made out. In order to discourage the
practice of retaining a lease without doing a fair amount
of yearly work upon it, a certain payment might be
agreed upon as the least that must be paid every year,
say a thousand dollars. That would also ensure the
government from loss in the expense of inspection and
underground surveys. Twenty years would seem to be
a suitable enough length of time for the leases to run, if
the present Kaitakushi have the right to alienate such
valuable public property for so long a time. Thirty cents
a ton would seem to be a fair enough royalty to exact in
Yesso, and the price of coal here in the East is so high
that such a rate would not discourage mining. In
order to prevent the mischief that might arise at any
future time from favoritism in the allotment of mining
claims, and to accrue to the government the profit that
should properly accrue from any great demand for such
privileges ; they might under conditions like those just
named be offered one by one to the highest bidder, to the
man who would pay the largest sum down for the lease.

It is unnecessary here to go into the matter more

particularly ; it will be time enough to work out the minute details whenever it shall be decided to adopt the general plan.

Kayanoma ; 2 July.—With regard to permitting foreigners to undertake mining in Yesso, it may be best to say a few words ; although no argument is needed to convince the inhabitants of most countries of the advantage of having the aid of foreign capital in the development of their natural resources. It is, to be sure, generally believed that there is some reluctance in the present case to allow foreigners to participate in such opportunities as may exist here for amassing wealth by mining enterprises ; and many even think that, if a foreigner should be permitted to embark in an industrial enterprise and should find it profitable, the jealousy of the government against him as an outsider would thereupon contrive such methods of hampering him as would result ultimately in his ruin. It is hardly to be supposed, however, that any government with the slightest pretension to enlightenment could be guilty of such childish folly as that.

But even if there be any political or other reason for excluding foreign capitalists from operating mines by themselves in a country to whose laws and courts they are not yet willing to submit themselves ; there would seem to be no reason for discouraging the association of foreigners with natives in such enterprises. On the contrary, there are very strong arguments in favor of the plan. Besides the addition of foreign capital to native, which would necessarily tend to increase the business, the wealth and the revenue of the country, there would be a great advantage to the native capitalists, hitherto inexperienced in modern business and perhaps timid about embarking in it alone. Associated in equal shares with foreigners well trained in the best business methods of western countries, they would without cost get the benefit of all their skill and experience. Hardly any better way could be devised for introducing into the country those excellent modern methods of business, that are the result of centuries of experiment and progress, and are so greatly lacking in Japan. The Japanese would be very greatly gainers ; but the foreigners, on the other hand, would be free from harassing suspicions as to the treatment they might at any time receive from the government ; for the very fact of equal association with natives would protect their capital and enterprise.

Sapporo ; 7 July.—On the 27th of June, Sunday, we rode from Kuromatsunai to Esoya, 15 miles (6 ri); and on the 21st from Esoya across Raiden Mountain to Iwanai, 17½ miles (7 ri). The road across the mountain is still extremely bad, especially on the Hakodate side. On approaching Iwanai we found half a dozen men making the very stony road a little smoother, and it seemed at first as if here at last was a place where the Kaitakushi was attending a little to road repairs ; but no, the villagers were themselves doing the work in preparation for a religious festival. It seemed for a moment almost a pity that superstition did not require the dragging of a heavy car across Raiden Mountain every year ; for in that case the road would be sure of being put in good order at least once a year. As it is, both the roads and the hotels of Yesso seem to have been gradually but steadily deteriorating for at least three or four years. The new road from Hakodate to Mori and Mororan to Sapporo and Otarunai has, to be sure, been built ; and a few new hotels of about the same date show that there were hopes then of speedy improvement in the condition of the country.

On the 29th of June it was rainy all day and we stayed at Iwanai. I wrote a portion of this report, and the copyist copied another part of it. On the 30th it was also very threatening weather all the morning, and I still worked at the report. In the afternoon, however, we rode from Iwanai to Kayanoma, 7½ miles (3 ri). In passing Horikap we noticed that the large building of the old abandoned salt manufactory had been turned to account at last as a place for sawing lumber and for a carpenter's shop.

On the 1st and 2nd of July we visited the coal mines. We found that a fault of about thirty feet had been met with in the Honshiki drift and had been overcome by a crooked tunnel through the rock, but that after following the bed less than fifty yards further it seemed to be growing thinner again as if near another similar fault. The Shinshiki drift had reached the first fault and stopped ; and the Midzunukishiki drift had on the Honshiki bed met a like thinning out, and probably likewise a fault near the entrance of the Honshiki, working south westerly. Working north-easterly it had not nearly reached the first mentioned fault, but had come to a rather thin place in the bed ; which was taken to be another fault, but no doubt corresponds to a thin place in the Honshiki drift where there was no fault. Owing to these faults and thin spots, the progress of the drifts had been stopped, and, as the coal opened up by them had become nearly exhausted, there was room for only nineteen men to work in the mine. The rest of the eighty miners who were there two years ago, had gone away to Otarunai, Esashi and other more or less distant places. The mines had therefore almost wholly lost the very great merit of being already so far opened up that a large number of miners could be at any time put into them, and a large quantity of coal taken out at very short notice. It would be necessary, as with a new mine, to drive the gangways forward a long distance, a slow process, before room could be found for many miners. There are, however, three or four gangways already begun, so that by punishing them all forward space for a pretty good number of men could be obtained in a somewhat shorter time than in a mine that is wholly new.

We visited the Tarioko drift that opens the Kosawa No. 1 coal, but it had also been abandoned because the bed had grown thinner after going about 80 feet towards the Honshiki. It is probable that there is a fault there, in addition to the one in the Midzunukishiki near the Honshiki, and there can now be no doubt that by means of these two and possibly other faults, amounting in all to about 70 feet, the Kosawa No. 1 bed will prove to be the same as the Honshiki bed. It becomes clear likewise that the Osawa No. 2 bed is the same as the Kosawa No. 1, which it closely resembles ; and probably there is no great amount of faulting between those two points. As Osawa No. 2 is at a level of 79 feet below the Tarioko drift, and is the lowest drainage level of the bed in the Kayanoma valley it would be very desirable to open it up by a drift, as recommended indeed in a former letter, where its identity with the Honshiki bed was already declared (although the faults that make the Kosawa No. 1 to be the same were not known.) Without waiting for that letter a new drift had been driven a few yards upon Osawa No. 1 in the belief that it was the Honshiki bed instead of the Midzunukishiki bed, which it really is. The coal in the new drift was about four feet thick, and was said to be of good quality ; so that it might be worth while to continue the drift, for the sake of working that bed, if it be not desired to lay out the whole available force for the present in working beds that are still thicker. As the Tarioko and the Midzunukishiki drifts are on the same level, it is hardly worth while to go on with the Tarioko, because the two faults would have to

be tunnelled across and very little coal obtained after all ;
and that little can be taken out by the Osawa No. 2
drift.

The coal is still carried to the harbor, they told us, on
the backs of horses, as it was last fall, but the inclined
planes, the tramroad and the cars were very soon to be
used again. With so costly a method of transportation it
must have been lucky on the whole that the yield of the
mine was so small.

Sapporo ; 8 July.—Late in the afternoon of the 2d of
July we left our quiet, cleanly quarters of Kayanoma and
returned to Iwanai, $7\frac{1}{2}$ miles (3 ri) ; myself on horseback,
the rest still later by boat. On the morning of the 3d,
just as we were setting out for Oshoro, $37\frac{1}{2}$ miles (15 ri),
Mr. Ichichi, now one of the Iwanai officials, came to
talk about a breakwater that it had been proposed
to build at Iwanai, partly for the benefit of the
coal trade. We spent an hour in walking and
rowing about to see the place, and of course the
knowledge to be gained in so short a time could only be
very general and derived chiefly from the well informed
inhabitant of the town who guided us and was evidently
strongly in hopes that a breakwater would be built.
Before actually beginning so expensive a work it would
be best to have a thorough survey of the place and of its
surroundings made, and, if possible, to take the advice of
somebody more specially conversant with such matters
than I am. The shore near the anchorage runs north-
westerly, bending round to westerly within a few hundred
yards. Just at the bend, and outside of the anchorage,
they say there is a rocky shoal 200 yards wide that runs
out nearly at right angles from the shore, reaching to a
depth of only 5 feet at 100 yards from shore, 10 feet
at 200 yards, and 18 feet at 400 yards. At the anchorage
the water is deep, even up to forty feet or so. Twenty

or thirty junks were at anchor there, close to each other, side by side in rows, at the time of our visit ; and there were a number there last October at the time of the storm that made so many wrecks at Furubira, Otarunai and other places of the west coast ; yet there were no wrecks at Iwanai. Nevertheless the bottom is only sand, not good for anchoring. The plan proposed was to build out on the shoal a wooden pier like that of Mori (which they said cost 30,000 rio). The bay is so very wide that only a small space would be thoroughly land-locked by such a breakwater, even if built 400 yards long ; but the protection to the whole of the present anchorage would be very greatly improved. From what I have seen of piers and breakwaters in similar extremely exposed positions, I am confident that such a structure as the one proposed would be broken almost every year by the enormous force of the waves striking broadside against it in heavy storms, even if built of cribwork filled with stones, much more solidly than the Mori pier. I learned at Mori that the *hiba* wood used for the pier was not attacked by the ship-worm and would last fifteen years in the water. It would be necessary however to count upon replacing it as often as that. It would therefore probably be much cheaper in the end to build with stone, which they said could be quarried at no very great distance. Owing to the shallowness, the amount of stone needed would not be very great, even if a very gentle slope should be given to the outer side of the breakwater. With the map of a proper survey it would be easy; o calculate the amount of stone needed and to calculate the cost of the breakwater, and so to judge of the feasibility of building it for the amount of shipping to which it would be likely to give protection. If the outer slope should be made very gentle and protected with large blocks of stone, perhaps partly with cut stone, it seems to me that little or nothing would afterwards ever be needed for repairs. There is no stream emptying into the anchorage to silt it up; and the unsheltered shore is quite distant, and for a number of miles merely a gently sloping sand beach without streams larger than small rivulets; and the entrance to the anchorage is its widest part; so that the waves and tides would hardly bring much sand to be deposited in its quiet water and to be dredged out again when the water became, in the course of many years, too shallow. As far as coal shipping is concerned, it seems to me that such a break water would on the whole be less feasible than the shorter one that has been proposed for Chatsunai and Shibui harbors, so much nearer to the mines, and so favorably situated in respect to the stone for building it, and to every thing else except the rocky bottom, bad for anchors (and in so small and well protected a harbor, the vessels could perhaps be tied almost always to posts on shore). Nevertheless, when the Horikap and upper Shiribets valleys are opened up by waggon roads and farms, the business of Iwanai as the natural outlet of a large and fertile region may very well justify the building of the proposed breakwater (in a prudent economical way); especially since harbors are so rare along the coast that a good one here and there would become of great importance as the resort of large vessels, which could take to Nippon or to China the fishery products brought from neighboring portions of the coast by small junks. The proposed Shibui breakwater, although making a harbor large enough for shipping the coal, at least by steamers, might not give protection to so much shipping as the business of Iwanai would require.

Just as I was mounting my horse, my opinion was asked as to the buying of a steamer (a tug, as I understood) for the coal business. It would certainly be worth while to have one in case an artificial harbor should be made and coal should be mined in any quantity ; but at present there seems to be no sufficient reason for incurring such an expense, with such a risk of loss by shipwreck.

I was finally off by eight o'clock and was surprised to find that my companions, and most of the baggage horses were equally late. The road as far Rubeshibe across the mountain was in very bad order a great part of the way and here, as elsewhere, repeated trips during the last two years have shown the deterioration to be pretty constant, and the repairs, within that time at least, to have been apparently none at all. From Rubeshibe I wished to take the new road to the sea beach near Oshoro which the landlord said last October was about finished and was to be opened within four or five days ; and by which the distance to Oshoro would be four ri instead of eight, without a hill as far as to the sea shore. I had been delighted to give the Kaitakushi great credit for making at least one such improvement in the wretched roads of Yesso ; but was afterwards told by a high Kaitakushi official that he had consented to the building of the road, but that the expense had been borne by the neighboring inhabitants, who had asked leave to make the improvement !

In answer to inquiries about the new road on approaching it, no very clear account of it could be got; and the impression was given that, although somewhat travelled over, it was not in very good condition. Thinking that if others had passed over it with horses we could do so too, and that, even though we went slowly, the saving of four ri was very important, I rode down it at four o'clock, after finding with great difficulty and delay the place where it began. It seemed but little travelled, although generally a pretty well marked track, and it turned out that no horse had ever been over the greater part of it before. Nevertheless after an adventurous trip and one or two hair breadth escapes from losing the horse by bad bridges I succeeded in reaching the sea long after dark, and then by an easy and mostly familiar road of nearly two ri arrived to Oshoro about half past eleven. Half an hour later, as it was beginning to rain, the groom and two other servants by the old road by way of Yoichi, arrived with some baggage and all the others had gone that way. The new road is still very far from being so short as had been said ; but makes the distance from Rubeshibe to the Yoichi ferry perhaps a ri shorter than the old road does (a distance which I had fully lost in searching for the beginning), and is much more level. It would seem very well worth while to lay out the very small sum of money needed to make the few bad bridges good, and if necessary to dig ditches along the road. At present there has been so little travel that there is not much mud except at the streams.

It rained heavily through the rest of the night and the early morning of the next day, 4th of July, Sunday. But as the rain grew trifling towards noon, and the rest of the baggage had come, we set out at eleven, and rode from Oshoro to Zenibako, 20 miles (8 ri). The road as far as Otarunai was very slippery with mud, and the rest of the way very rough indeed with stones, so as to make progress very slow with an unshod, tenderfooted horse ; but we arrived in nearly six hours. It is this heavy tax by bad roads upon the time and strength of every traveller and every packhorse, that makes it so economical to the public in the end to build good roads and to keep them in good repair. About nine o'clock the cook and groom arrived on foot, and said that although all the rest had arrived safely at Otarunai, it had been impossible in two hours to obtain even three packhorses to bring forward the most necessary baggage. Such a fact seemed noteworthy in a town of 500 dwelling houses, especially as in a year and a half's travelling in Yesso no serious difficulty had ever been found in getting a much larger number of horses at shorter notice in even very small places.

The next morning, 5th of July, I rode from Zenibako to Sapporo, 12½ miles (5 ri); and the rest also arrived early in the afternoon.

Sapporo; 11th July.—The improvement of the road from Zenibako to Sapporo within the last two years is very marked, especially in the number of new houses and farms; tending to confirm what I have repeatedly insisted on, that good roads are what is most essential for inducing immigrants to occupy the many fine farming regions of Yesso. The apparent prosperity along the road was, however, due in part perhaps to the new immigrants brought lately by the Kaitakushi and settled in a neat looking village to the left of the road before reaching Sapporo. I have already in last year's report argued against the policy of bringing immigrants to Yesso by so costly a method, when the same expense laid out in roads, schools and such public improvements would probably attract a much larger, wholly voluntary, immigration, which would entail no unusual responsibilities upon the Kaitakushi.

The village seems to have been built with its houses very close together as has been the old custom in countries infested with savages or disturbed with guerilla warfare, instead of having each house upon its own farm as commonly happens in the new western states of America, and as seems more suitable to the peaceable circumstances of Yesso. There are no fierce savages nor lawless brigands here and in case of foreign invasion this wooden village would give little if any better protection against modern artillery than scattered homesteads would. On the other hand the farmers must now walk daily a long distance to and from their fields, and by a waste of much labor bring the produce to the village. Moreover in the driving out of troublesome wild beasts from any region, scattered habitations all over it must be a most effective agency, as compared with isolated villages with wide spaces between, as may be seen by looking at thinly inhabited portions of America with very few such animals, and at India with a very large population concentrated in villages and with great numbers of beasts of prey between. Though the wide intermediate fields be frequently visited, the wild animal does not always find them occupied and can frequently pass through them undisturbed ; but if houses be scattered there, they are always there and he always meets with men, his haunts and his freedom are effectually broken up, and if he escapes with his life he must remove to some wilder region. The advantages of a closely built village are that the children can go readily to school every day, the militia men can assemble easily for their occasional training, and, in the present case, the government supplies of rice can every month be conveniently distributed. But farms here are so small that even with scattered houses the children would not have to go too far, especially as their work at school is not bodily labor, and a walk before and after would do no harm ; and the trainings and the distributions of food are so much rarer than the farmer's daily visits to his field as to deserve much less to be considered in planning a settlement.

The village in question is some distance back from the main road like a number of other farming villages around Sapporo that are quite hidden from the sight of travellers over the main highways. The villages thereby lose in a great measure the benefit of the outlay that has been made upon the great road, and the country continues to look wild and unattractive to a new immigrant. Even the more favorably situated lands along thoroughfares must therefore wait long before they will increase in market value and be much sought for. The earliest comers, too, lose the natural reward of their being the first immigrants, and have their hardships increased, and must consequently give to the friends they left behind a less pleasing picture of their experiences and hold up to them much smaller inducements to immigrate likewise. It seems therefore a very bad plan for the Kaitakushi to reserve for future immigrants and perhaps higher prices the better situated lands; for it must be the aim of the government to have the country well settled up and consequently productive, and a few dollars more or less in the price of the land is of much less importance. However desirable it may be that government officials should become personally interested in the country by owning land there, any tendency on their part or on that of their friends to retard the progress of settlement, by buying up the best tracts of lands and letting them lie idle until they can be sold at high prices, should be severely discouraged. A yearly tax levied upon land with some reference to its natural value, whether cultivated or not, would be a discouragement to holding lands idle for a long time.

Sapporo; 13 July.—Perhaps indeed the best method of raising revenue from land would be to divide all the land of the country permanently into four or five classes according to its value; as, for example, good farming land, mountain land, swampland, townlots, and land bordering on a great thoroughfare, say a river, the sea, or a road; and to have them always thereafter taxed each class at its own rate, but with that rate bearing a certain permanent relation to a standard that could be fixed every year, or more seldom, according to the needs of the government. In that way the excessive labor and almost inevitable errors of a minute valuation (such as is undertaken from time to time in some parts of India) would be got rid of; and yet substantial justice would be done. Whenever a farming region is opened up by new roads the lands adjoining them could easily be advanced from one class to another; and likewise in the case of building new towns. As compared with the Japanese practice of taking yearly a certain portion of the product of the land, it would greatly discourage letting land lie idle and therefore unproductive not only to the revenue but to the general wealth of the community, and in like manner would discourage inefficient and harmful methods of cultivation. Farmers would be encouraged to lay out money in permanent improvements of the cultivable land, because, although they would thereby obtain larger harvests, their taxes would not at the same time be increased. Money so invested would in fact become in a manner exempt from future taxes; but the community would gain correspondingly in its general wealth and prosperity, and its greater ability to raise the standard rate of land taxes. As new lands are taken up, either the land revenue could be increased, or the standard rate could be lowered.—In Yesso, however, little or no revenue seems as yet to be raised from the land, and perhaps for the encouragement of immigration it is best that it should continue so. Still, enough might be raised from land owners to keep the roads in repair, if not to build new ones. But if the horses of Yesso should become the property of the inhabitants, instead of that of the Kaitakushi almost exclusively, the road repairs and even some at least of the road building could be done with equitable and easily borne taxes on beasts of burden and vehicles, which use and wear out the roads. By an extension of the same principle the schools, hospitals almshouses, courts, police, and prisons could be supported by taxes on wines, tobacco and perhaps some other luxuries ; the land survey and military expenses by a land tax ; the coast survey, light-house and naval expenses by a tax on shipping or on fishery privileges ; and the geological survey and mine inspection expenses by the coal royalty. Under a system of that kind the connection between a tax and the use to which it is applied would be so obvious that even direct taxes would cause no more discontent than ordinary indirect ones.

Sapporo ; 15 July.—As I have been some days in Sapporo, it may be worth while to comment on a few things that have come under my eye, though they be out of my special line and no official information in regard to them has been given me.

The streets have a decidedly busier look than they had last year, although the still very numerous empty and abandoned houses are now much more striking with their tattered window paper and dilapidated walls and roofs. No doubt the new settlers in the neighborhood and the farms and mills have added to the business of the town ; and the present improvement is apparently better founded, more cautious and more likely to continue than that of three years ago, which must have been based on very extravagant anticipations. There is some reason to hope that with the present steady advance all the empty houses may before very long be permanently occupied.

The chief new enterprise now going on in the town itself seems to be the building of the two or three large new houses of the boys' and girls' schools. The boys' school is pretty well situated in the quiet, airy quarter that has been set apart for the homes of officials and for offices. The site, however, seems to be none too well provided with space for a large play ground, so essential not only for the amusement, but for the health and strength of such a large number of children, no matter of which sex. Unoccupied land is not so rare in this region as to give any excuse for neglecting a matter of such importance. But the girls' school is far worse off than the boys' in that respect, and could indeed hardly have found a more unsuitable place than the one it has, which was, however, exactly the right one for the old honjin that lately stood there. It is the very middle of the business part of the town, on the busiest, noisiest thoroughfare, and with scarcely a foot of land near it that can possibly be made into a play-ground. Indeed only one worse place in the whole town could have been chosen, but that, although created by the Kaitakushi, and, as I understand, peopled by it, and less than a quarter of a mile from this home for young girls, would in almost any other country be thought too disgraceful even to be mentioned. The school buildings themselves seem large enough to give room for the lodgings of a pretty good number of scholars.

Sapporo; 16 July.—It seems however to be thought nowadays by those who are best informed on such matters that it is not a good plan to shut up numerous individuals of a single class in one establishment, and that large school dormitories as well as large hospitals or large prisons tend to create and propagate great evils that tend on the other hand rather to disappear if those afflicted with them be scattered widely among a population that is free from such taints. In other words it would be much better for the physical, moral and intellectual health of the scholars to live here and there among the respectable families of the town, a very small number at most in each house; where they would have a much larger share of home influences (so essential to the formation of good character in children) than is possible in a large dormitory. Certain very grave evils that are apt to result from assembling so many youths together in one establishment have hitherto happily been so rare in Japan as hardly even to be dreamed of. It may seem useless to talk of a family system in a town where at present scarcely one official's wife makes her home ; but the dormitories could at least have been divided into much smaller separate ones that should have some respectable family or individual to take charge of each, with a view to a still more complete subdivision as soon as practicable.

It is clear, however, that the scholars are not children of inhabitants of Sapporo, even · if they have parents at more or less distant places in Yesso. It is right then to consider whether there are any advantages in having the schools here rather than in Yedo where they have been hitherto. If the expense of lodging and boarding and properly teaching the scholars be less, and the social and intellectual influences better here than in Yedo, there can be no question about the propriety of the removal ; for, although the climate of Yedo be probably the best in the whole world, that of Sapporo is sufficiently good and healthful, delightful in summer and not by any means excessively rigorous in winter, in spite of the bad reputation it has among the highly favored inhabitants of Nippon. But it can scarcely be doubted that Yedo can claim superiority in those other respects even more than in its climate. Especially the additional difficulty and expense of obtaining here permanent satisfactory instruction would alone be reason enough for not making the removal.

Moreover, a foreigner accustomed to see how the different departments of a government in any foreign country cooperate one with another cannot help asking why these schools, chiefly for the training of very young children with foreign instruction to become teachers at some future time, should be kept by the Kaitakushi at all. Why should the children not be sent as Kaitakushi pupils to the schools of the Educational Department of Japan? or why should not teachers for the local schools of Yesso be taken, when wanted, from among the graduates of the normal schools of the Japanese government? As regards economy and efficiency one consolidated management of schools for the whole empire would seem likely to be decidedly better than the present arrangement. Of course the greatest need of Yesso is primary schools for all its children, both Japanese and Aino, children that are still too young to leave the homes of their parents; and it is of the utmost importance to find suitable and numerous teachers for such schools as soon as possible. The pupils for higher schools are much fewer, and their parents can generally better afford to provide for their education themselves.

Among the public works going on near Sapporo is one that deserves mention as it seems to endanger the town itself; I mean the dam that is building near the upper head gate of the canal that runs through the village and supplies the saw-mill. The head gate and its embankment were carried away by a freshet late in 1873, but early in 1874 were rebuilt much better with the embankment higher and larger, and are now apparently still improved by the consolidation that has come with time. But as the gate is just at the outer edge of a bend in the side channel that feeds the canal, it must in times of high water be fully exposed to a current of tremendous force. It is clear that for such an emergency the channel below the gate, as far as to the main stream close by, should be kept as free from obstructions as possible, in order to favor the harmless escape of the swollen waters. It is therefore a matter of no little astonishment to find that laborers are building a dam across at that very point, just below the gate. The dam, to be sure, is not so high as the gate and embankment, but must be at least four or

five feet high, and will tend very greatly to hinder the
flow of the water even in time of flood. It will tend to
raise the level of the water, so that in the highest
freshets it may perhaps run over the rather narrow
embankment and very likely loosen and wear away the
earth of which it is chiefly made, especially if the logs
that form its cribwork or sides shall in the mean time
have lost some of their strength by decay. The water
will be aided in such work by the huge drift wood that
the current will send with great force against the embankment in sweeping round the curve. The reason for
building the dam has probably been that the lower end
of the channel had become worn so low that the water
could hardly enter the gate. But a better remedy would
have been either lowering the bottom of the gate (and the
steep grade of the canal would admit of deepening its
upper end); or else a temporary very low dam of loose
materials (such as now enables the dam to be built without inconvenience from water) could every year be made
at very small expense, and allowed to be swept away by
the freshets; or at worst a substantial dam could be built
in the present place, but very low indeed, so as to
obstruct the channel as little as possible. But even a low
dam, though in a less degree than the present high and
very strong one, would not only tend of itself to raise the
water, but would tend to cause the accumulation of drift
wood that would give the effect of increased height to the
dam. If the embankment should give way in a time of
very high flood, such as happens once in several years,
the disaster that would be wrought in the town below
would be extremely great.

The new bridge across the Toyohira here is another
case where the intention was better than the result.
They say that the longer span fell down of its own weight
and that the shorter one can barely hold itself up. The

highest proof of wisdom is said to be a consciousness of one's own ignorance ; but, though high wisdom be rare in other countries, it must be difficult to find one of them where the building of a bridge of such size could be entered upon without at least taking and following the advice of a competent civil engineer specially trained for such work. Many communities, corporations or individuals with nearly two million dollars a year to lay out in public improvements, a good part of which should be engineering works, such as roads and bridges, would perceive the economy of keeping in permanent employment at least one good civil engineer, who could be trusted not only to give advice about such works, but to take charge of building them when once determined on. It might be gratifying to a foolish vanity to do everything one's self with, or without, or even against the advice of a man specially trained for the work, but it would be rather an expensive way for Japan to learn the principles of engineering.—Meantime the little bridge near Hakodate has still the honor of being the only finished bridge in Yesso that can have cost more than a hundred dollars.

Sapporo : 17 July.—The Sapporo mills, however, are an establishment that the Kaitakushi may well look on with some pride ; and indeed their merits seem now to be better appreciated than they were a year or two ago, before the works were fairly in running order, and now very active operations are kept up daily from early morning until night, instead of the half time of former years. All parts of the wood working machinery seem to be in use : the steam saw-mill, the water saw-mill, the planing machine, the tonguing and grooving machines the scroll saw, the shingle machines, the lath machine, all kept busy by workmen who are now familiar with their management. Besides that, handsomely finished doors,

window sashes, chairs and other objects are turned out in
large numbers by skilful workmen, who use moulding
planes that were made here. The large number
of logs that have unfortunately been kept waiting
in the ponds too long, are now getting sawed up, and a
good store of boards, planks, slitwork, shingles, and
other sawn material is accumulating, and, by seasoning,
becoming far more valuable than the green wood of which
the first houses of Sapporo were built. One very curious
circumstance, however, is worth mentioning, and that is,
that in the midst of all this active use of the best modern
wood working machinery, there is a long shed in which
25 or 30 stalwart men are just as busily sawing large
logs into boards by means of the primitive hand saw;
and that, too, at such wages that they can earn about 70
cents a day, or half a cent a foot of board measure. If
even hand sawing at such wages is profitable here, the
power sawing must be much more so.

It seems indeed quite unquestionable that such mills in
private hands would give a handsome profit; for even
with the high wages of the western coast of America
lumber is sent many thousand miles right past Yesso to
the Chinese market. What has been said about the
working of coal mines by the government applies almost
equally to its sawing of lumber. But as the capital
needed for a saw-mill is much less than for a mine, it may
have been very well worth while for the government
to show by an actual experiment what could be done
by small capitalists, and to put up mills of its own
as a sort of training school for workmen. It must, however, in other respects be most desirable to encourage
private parties, especially Japanese associated with
foreigners, as urged above in regard to the mines, to set
up mills of their own, with the privilege of cutting timber
within a certain district on condition of paying a moderate

royalty for every thousand feet of board measure. It would be necessary, of course, by forest inspection to keep some oversight of the cutting of trees, in order to prevent a wanton waste of what must become more and more valuable material.

It is not to be supposed that the supply of timber is literally inexhaustible, though it is known to be very great. No estimate of the whole amount has ever been made, and scarcely a measurement has been made on which a general estimate could be based. Still, as even a very roughly approximate idea is better than no definite one at all, it may be worth while to consider the matter here. It would seem that while the trees of the Ishcari valley bottom are large and good, those of most of the rest of the island are in general of no very great value. Last summer in showing the assistants at Poronai how to measure a forest, I made the only such measurement that has been made in Yesso. The spot happened not to be very heavily wooded, but may, perhaps without enormous error, be taken as an average of the forests of the bottom land of the Ishcari valley. We found about 21,000 feet of board measure to the acre, or say 13,000,000 feet to the square mile. The extent of the Ishcari bottom lands has been very roughly estimated at some 600 square miles; and if we reckon that one half of it be covered with forest, aside from the swampy and prairie lands, we have 300 square miles of good forest, which would contain, say, 3,900 million feet of timber. A steam saw-mill like the one at Sapporo could cut, they say, 10,000 feet a day with ease, or say three million feet a year. It would take thirteen such mills then at least a hundred years to saw up the timber now standing in the valley; and by that time the trees that are now small would have become large, so that the supply of timber would be almost as great as it is now. As the bottom lands are, however, very good indeed for farms, it will no doubt be thought best to cut the forests quite off and to depend on the abundant mountain lands for future supplies of timber to the island, if not for export. It would take, then, five times thirteen or 65 mills to saw up the timber in twenty years; or allowing for the growth of the trees in that time, say eighty mills would make a clean sweep of the forests in twenty years. About 5,000 million feet would have been sawed, which at a royalty of one dollar a thousand feet would have yielded a revenue of five million dollars with very little expense for collection and inspection, and the land would be very greatly improved for farming. But it must not be forgotten that the estimate is an extremely rough one; and although the best and heaviest timber is generally near the streams, and the swamps and prairies further back, yet the difficulty of getting the logs to the nearest water that could float them would perhaps very greatly lessen the number that could profitably be brought to the mill and to market. The fine large sticks of timber would also be a comparatively small part of the whole.

It is readily to be seen from the estimate, rough as it is, how insignificant is the lumber interest of Yesso compared with that of the coal. Probably the coal of the little Kayanoma field alone would yield within an easy depth, at 30 cents a ton, a greater revenue than all the lumber we have been reckoning on.

However unnecessary it may be for the government to embark in risky industrial enterprises like mines or even saw-mills for the sake of showing capitalists what may be done and how to do it, the objection has much less force when brought against government farms. For the government can do much in the way of teaching the small farmers, mostly men without capital, and in trying for them experiments in cultivation and in the introduction of new and useful plants and labor-saving implements. The Kaitakushi has already done admirable work in the bringing of cattle, fruit trees, grain and other plants from abroad and in their abundant propagation and distribution. In a few years the new orchards alone, amounting to hundreds of thousands of trees in bearing, will bring the good fame of the Kaitakushi most agreeably into every body's mouth ; and will excuse many a mistake that may have been made.

The government farms at Sapporo have been very much enlarged since last summer, and it is to be hoped therefore that they are not a costly burden to the department, as indeed they ought not to be. Although their main object is not the earning of a profit to the government yet they ought to be managed with strict economy, and give a wholesome lesson and example to the farming public in this point as well as in others.

Among the most important things to teach the farmers is how to substitute the cheap labor of animals for that of men and how to make one horse take the place of ten in this region of wide, level, fertile land that must be so low in price that even small owners ought to be able with proper appliances to cultivate larger fields than in Nippon. The Kaitakushi has already done something in the way of bringing ploughs, mowing machines, waggons and carts into use on its own farms, but much yet remains to be done. Strong waggons and carts that can be used even on rough roads seem to be among the most useful objects for the inhabitants of the country, especially if the good roads become more numerous. At the Sapporo mills it must now be easy to build one horse carts that could be

sold at a moderate price to the farmers and save them the
maintenance of many of their horses besides saving a great
deal of their own time and labor.   But the only waggons
and carts that I have happened to see here or at the farms
are of poor patterns for general use.

Two or three things of minor importance have happened to come under my eye and perhaps deserve a word.
—The dwelling houses in the official quarter of Sapporo,
though built neatly in foreign style, have a singular
appearance from the lack of chimneys, and even the new
school-houses have none.   In the close, unventilated
rooms of such houses the Japanese brazier would be a
most dangerous heating apparatus ; and, even if enough
outer air be admitted to prevent suffocation, yet the
tendency to lung complaints must be very much increased.
The custom adopted for some houses of putting an un-
sightly stove pipe through a hole in the outer wall is a
very unsafe one.—For foreign houses with high windows
and without Japanese mats foreign furniture, chairs,
tables and bed steads are very necessary, but seem to be
still extremely rare in Sapporo, though they can readily
be manufactured at the mills.   It is perhaps to be re-
gretted that so very radical a change in the style of
architecture should have been attempted instead of trying
to adapt the ordinary Japanese house to the winter
climate.   Even a few dollars laid out in replacing most
of the wood of the rain doors and the paper of the
windows with glass and a little more attention to closeness in fitting the slides would go very far towards
accomplishing the end in view, and in fact would need
scarcely anything further except chimneys, in case wood
be burnt.   Complete tightness in the fitting of doors and
windows is no object, for it would have to be made up for
by special ventilators.—The principal streets of Sapporo

have lately been metalled with river gravel, a very great improvement in rainy weather on the slippery black mud that the rich soil naturally produces. But the improvement would have been still far greater with scarcely any increase of expense, if the gravel had first been thrown upon large inclined screens so as to separate the sand and smaller pebbles from the very large stones that now encumber the road every where and cause the "improved" portion to be carefully avoided by all passers. Even waggons and iron shod horses would need almost an eternity to pound such material into something satisfactory.—The new post and rail fences at the government farms have been made with almost double the labor that is necessary ; for the posts are the chief source of outlay in such a (three rail) fence and one half of them here are quite useless, those, namely, that are in the middle of the rails.

Sapporo: 19 July.—Since our arrival in Sapporo the assistants and I have been occupied daily with office work, report writing and map making, so far as the extremely poor accommodations for mapping would permit. We have now finished the report and map of the Makumbets Coal Field, and the report shall be sent with this. For greater safety the map would better wait until our return to Yedo ; but, as I find there is a good photographer here, perhaps it will be convenient to send soon a photograph on half the original scale.

We are all impatient to get into the field now ; but owing to the breaks in the telegraph have been some days vainly expecting to hear definitely your wishes in regard to the survey for a coal road from Poronai, on the whole the best place to begin working the great Ishcari coal field, to Horumuibuto, the nearest practicable shipping point on the river ; since there was some desire expressed for such a survey on my arrival here.

Sapporo ; 20 July, 1875.—I improve the opportunity of sending you now this portion of the report of the season's work ; and have the honor to be,

Sir,

Your most obedient servant,

BENJ. SMITH LYMAN,

Chief Geologist and Mining Engineer to the Kaitakushi.

Camp near Bibaibuto ; 28 July, 1875.

His Excellency K. Kuroda,

Kaitakuchokuwan.

Sir :—On the 20th of July, (the day I sent to you by
mail my report of progress up to that date and the report
on the Makumbets Coal Field) my assistants and I went
to look at the bog iron ore places near Hiragishimura, and
about 3 miles (1¼ ri) south of Sapporo. Messrs. Kuwada
and Mayeda, of the assistants, had gone there with some
laborers the day before to attend to any digging that
might be necessary in order to expose the ore properly ;
and had brought back sketches from a little rough
surveying and measurements of the ore.

Hiragishimura is a farming settlement, a mile or more
long on either side of a wide highway, through the middle
of which water runs in a narrow canal northward towards the Toyohira from a branch of the same
stream, called Fastingman River (Shôjin-gawa) because
fish are kept out of it by a weir. Some half a mile from
the southern end of the village is the nearer ore place in
the side and bottom of the canal and in the midst of a
swampy forest. Owing to rain that had fallen heavily in
the night and swollen the streams we could only see a
very few inches of the top of the ore; but Mr. Kuwada
had carefully measured it the day before and found it 2½
feet thick with perhaps a foot of reddish brown loam
above it and 2½ feet of soil above that ; making about
six feet in all for the depth of the hole, which had at
bottom below the ore, sandy clay with small stones.

About 15 yards northerly along the canal, ore is exposed again by the upturned roots of a fallen tree ; but no digging has yet been made there. Indeed, for about 50 yards northerly from the digging, small ore bits are to be seen among the earth that has been thrown out in digging the canal ; but at 70 yards in that direction no more of them are found. In like manner to the south, ore bits are numerous for 50 yards but altogether disappear within 75 yards. The ore is a hard brown hematite, but very porous and therefore apparently light in weight ; and a rumor has reached me that in an analysis by Mr. Ishibashi it yielded about 56 per cent. of iron and a large amount of phosphorus. I know nothing whatever of his trustworthiness as a chemist, but, as I mentioned already last fall, it is highly probable that the amount of phosphorus would be large, as in other bog iron ore, and the quality therefore very bad.

It is not easy to ascertain the whole quantity of the ore ; but it is clear that it must be quite small. It seems to have been formed in a swamp (perhaps out of some small deposit of magnetic iron sand from volcanic rocks, for numerous bits of pumice and other volcanic stones are found in the alluvium here) and to lie nearly parallel to the present surface of the ground, which is still somewhat boggy. Without digging numerous deep holes here and there throughout the swamp it is impossible to tell with much precision the extent of the bed of ore or its thickness at all points ; but the canal happily supplies in a measure the place of so troublesome an exploration. For it has crossed the deposit in nearly a straight line, and there is no improbability in supposing the extent and thickness to be as great in that direction as in any other ; they are in fact just as likely to be greater as to be less.

It may therefore be supposed that the bed is only about a hundred yards in diameter, and at its centre $2\frac{1}{2}$ feet thick, thinning out to nothing at the edges all round. That would give perhaps 7,000 cubic feet of ore, or (reckoning such porous ore at twice the weight of water to the cubic foot) say 400 tons. Although the estimate is an extremely rough one it is enough to show that the amount of ore is unquestionably too small to deserve any further consideration.

There is another exposure of like ore near a clog-maker's house about a mile (half a ri) further south ; but as the common road to it was that day impassable on account of the wetness of the swamp, we tried in vain for three hours to find the place by some other way, wandering about hither and thither through the bog with a guide or two ; and finally some time after rain had begun to fall again gave up the search for the present as almost certainly not worth while. Mr. Kuwada had the day before made the most needful measurements. He found the ore in a hole dug in a flat field, about a hundred feet south-westerly from the house and from the Shojin-gawa, which runs here south-easterly. The ore was about three feet from the surface of the ground, and was lens-shaped in section, having a thickness of six-tenths of a foot at the middle and thinning out to a knife edge at the sides of the hole. Above the ore was perhaps a foot of dark brown loam, and above that, soil ; below the ore was perhaps a foot of yellowish clay, and below that sandy clay. The amount of ore here seems to be still so very much less than at the other place, as to be undoubtedly quite insignificant. If the bed had any extent, it would doubtless have been seen in the bank or bed of the stream close by.

On the 21st of July as the weather was still raining we kept on with our office work of map-making, instead of going to look at the bog iron ore near Bannaguro and Ishcari. Towards night word came that you had requested by telegraph that I should make the survey for a coal road from Poronai (or Ichikishiri) to the Ishcari river, as had been proposed to me some days before. The visit to the iron ore places (from which so little in quantity as well as quality of ore now seems likely to result) was therefore put off until some other drier and more convenient time ; and preparations for departure up the Ishcari were begun at once.

The 22nd of July, we tested our prismatic compasses by the compass of one of the transits and made out a letter of instructions for Messrs. Kuwada, Misawa, Kada, Mayeda and Nishiyama (who go at once to continue the survey of last season northerly from the Bibai valley towards the Sorachi, while the others accompany me at first on the road survey.) The rest of the day was taken up with packing baggage and with other preparations for the journey.

On the 23rd of July, Messrs. Inagaki, Takahashi, Saka, Shimada, Yamagiwa and myself set out in log canoes from Karekimura on the Toyohira, a league from Sapporo. I succeeded in camping that night on the Ishcari within a league and a half of Horumni.

At the Karekimura farms I was struck with the fact that the growing millet which was so abundant last year and year before, was now wholly wanting and replaced by barley. On inquiry I was told that barley had been found to yield better than millet ; and that this year a little wheat had been sown.

The next forenoon (24 July) we pitched our camp on the bank of the Ishcari a couple of hundred yards above the mouth of the Horumni after a journey of only about 4 miles (1½ ri) ; and walked about a little in the woods to see how the land lay. In the afternoon we began the running of a straight trial line towards Poronai as nearly as I could guess from the best mapping we had. In the evening, however, Mr. Kada came (for Mr. Kuwada's party had that day left Sapporo and had encamped a short distance below us); and gave a most important correction to the information we had hitherto had about the position of Poronai. The assistants who ran the line last summer from Poronai to the Ishcari had been greatly mistaken at Yedo in pointing out the place where they had reached the Ishcari. Mr. Kada had among the laborers with his party met with one who was with the party that cut the line, a man who was familiar with the Ishcari River; and had learned from him much more exactly the point where the line cut the river. Another Japanese laborer in his party and an Aino in ours confirmed the correction, and the next morning (25 July, Sunday) when the laborers, who had been hurried up by a message, arrived, we were able to lay the point down on the map pretty exactly. The position of Poronai on the map was thereby removed nearly two leagues from Horumuibuto, and Bibaibuto became nearer than Horumuibuto by a league or more in a straight line from the Ishcari to the coal. An intelligent, gray-bearded Aino in our party, who had lived all his life in the lower Ishcari valley and had the reputation of being unusually well acquainted with all its features, went with me to an open swamp near our camp, and pointed out Poronai mountain and Ichikishiri valley in the distance in a way that corresponded with the new correction of the map. To go straight to Poronai would require crossing the swamp where it was wide; to go round would make the road still longer. He said that straight across from near Bibaibuto there was no swamp; so it seemed better to try that decidedly shorter route on firm ground, in spite of the greater length of river navigation required. He said moreover that up to Bibaidap the river was not shallower than it is at a place near the mouth of the Ebets, and that it was about ten feet at both places in low water ; and that corresponded well enough with our soundings of last year. It was therefore decided to try a line across to Poronai nearly due east from just below Bibaibuto ; and we arrived here the same evening, having come from Horumuibuto about 13 miles (5¼ ri).

The Ishcari river at Horumuibuto is said to vary fifteen feet between low water (commonly in June and July) to high water ; and at the time of our visit there to have been about five feet above low water mark. A new stake there, with the feet marked on it, was observed by some one of our party to give the depth of the water as two feet.

On the 26th of July, we began a line about due east (S. 85° E. magnetic) from a point where the river runs due south. We had gone but 800 feet when we came to a large open swamp and could see distant mountains before us. On consulting our Aino who was so familiar with the region, in order to aim the more exactly at the nearest practicable entrance to Poronai valley (which we had not exactly on any map), he pointed very positively full 30 degrees to the south, and was sure he was not mistaken. So we began a new line in that direction from the point where the river ran at right angles to it ; but after going more than half a mile, all but the first 800 feet in the open swampy ground, we came to a large pond without outlet, which we were able however to get past by an offset of 400 feet to our right. Wishing to correct our course, if necessary, before going further, we consulted the Aino again ; and though he maintained at first that we were right, he admitted after a while that he had been mistaken and that Poronai was almost exactly in the direction our first line was aimed. The incident strongly confirms what I wrote last year against the infallibitity of the Ainos as guides and of their " sense of direction " which some travellers have been inclined to extol.

We returned therefore to the first line, and continued it across the swamp about a mile ; after changing its course to S 86°25′ E in order to touch just the foot of the nearest low hill that we must pass round to enter the Ikushibets valley below Poronai, as near as we could judge by our glasses. The swamp is by no means impassable on foot, although very wet and often ankle deep with water ; but I greatly fear it may prove a serious obstacle to the building of a coal road. Near the middle of the swamp one of the laborers easily thrust down a slender, iron-shod, bamboo surveying rod, ten feet long, full length without meeting anything hard except a slightly hard spot about two feet deep. Nevertheless the slope of the surface, about seven feet in some 3,000 feet from the river bank, which is itself about nine feet high, seems quite enough to drain the surface easily by shallow ditches which would not be long if dug to the river. If the whole or a large portion of the swamp should be drained in that way it seems likely that the ground would become firm, as it is now along the water courses, wherever there are any. Perhaps for a road it would even only be necessary to dig one good ditch on each side the whole length. The experience already had in digging the canal through the swamp north of Sapporo and the one near Hiragishimura may be a guide as to what may be obtained in the way of firmness in the present case.

The line finally adopted runs about 400 feet north of the northern end of the large pond already mentioned, which is without visible outlet and near the highest part of the swamp and is surrounded by a fringe of trees with rather firmer ground, although the pond is nearly brimful of water. The shores are somewhat curved, and the eastern and western ones about parallel, and perhaps 2CO yards apart, with the hollow of the curve towards the west. It seems at first a little puzzling to account for such a lake in the midst of so wide and nearly level an alluvial plain ; but it has struck me that it may be a portion of what was once a part of the course of the Ishcari. If that portion had been hollowed out unusually deep by a strong current around a bend, it would take longer than the rest of the plain to fill up to the present level, by the widely uniform deposits of mud in times of high water.

The Ainos say that the Kawanai and some other small streams, a short distance north of where our line will run, flow southerly into this same wide plain, and end there in a swamp or pond that has no outlet, a remarkable circumstance in this region that is so far from arid, especially as the Ikushibets is so near on the south and the Ishcari and and Bibai on the west. The small streams in question are also said to have few fish and no masu (salmon-trout ?) in them ; but the water of the ponds without outlet would not seem to be salt.

On the 27th of July, (yesterday) and to-day the weather has been very rainy indeed, so that we have had to keep in our tents and do merely office work, report-writing and a little mapping. While I am running the main line with the transit, one of the assistants levels it with a hand level, and two others attend to the chaining and rod holding. The two others have both here and at Horumuibuto made a survey with the prismatic compass of the river shore above and below our starting point.

Camp, Bibaibuto ; 29 July.—The clouds cleared away last evening for a little while just at the right time to observe the north star in order to determine the true meridian ; and as it was not later than the convenient hour of half past ten, we made the trial, and had a very satisfactory observation with a good transit theodolite. This morning by reckoning 16 compass readings (in two sets of eight each with the same result for the mean of each set) sighting in the direction of the star and of our first line from the river bank, and with the telescope sometimes reversed, the mean variation of the needle proves to be 4°32' west. The azimuth of the star as seen at the greatest eastern elongation was calculated to be 1°51' at this time and place. The variations of the different compass readings from the mean were for the first eight sights,-6,-5,-1, 0, 0, 4, 4, 4 ; and for the second eight,-5,-1,0-,0-,0-,4-,4-,4-.—The variation of the compass is given on Mr. Wasson's Ishicari River Map as five degrees west in 1873 ; and was found by Mr. Munroe to be the same at the Makumbets Coal Field last September ; but I am quite uninformed as to the special circumstances of their observations, instruments and computations. If no error has crept into either of the three determinations, the variation is decreasing very rapidly year by year and increases rapidly as you go eastward.

The weather to-day has been very threatening and the swamp is said to be so deep with water as to be impassable. We have therefore stayed in camp or near it, and have done some office work, besides finishing by daylight some of the observations and computations for the determination of the meridian and continuing to the river the the corrected course of the line we ran across the Swamp.

Towards night I made a canoe trip of an hour up the Bibai River and sketched its course, which was in general south south-easterly.

Tonebetsbuto, 12th September.—On the 30th of July although the weather was still very threatening and the swamp very wet, I went on an exploring excursion with with two Ainos, leaving the assistants to go on with any office work that might be yet unfinished. The Ainos and I went down the Ishcari in a canoe for a couple of miles to the first bend below Bibaidap, and then went across on foot about S. 74 degrees E. by hand compass about four miles ($1\frac{1}{2}$ ri) to the Ikushibets. I was still in hopes of finding that the old Aino guide had told me rightly at Horumuibuto that near Bibaidap there was a place where you could cross from the Ishcari to the Ikushibets without any swamp ; but this day's excursion clearly proved that it was not so. We had gone hardly 300 yards from the Ishcari before we came to an open swamp which reached to within a few hundred yards of the Ikushibets. Half way across we could see that it was a part of the asme swamp that we had entered in surveying near Bibaibuto ; and that it reached much further northward and very far southward, apparently as far as to the open swampy ground we had seen at Horumuibuto. In crossing we met with two large ponds one near either river, but did not have to turn far aside to avoid them. The pond water was undrinkable from the great amount of vegetable matter in it. Towards the middle of the swamp the ground was a little higher and firmer, and the grass very low to that the walking was comparatively easy ; but towards the Ikushibets side there was much more water and the bushes were higher and higher, so as to make it very troublesome to get along. Towards the north-east along the Ikushibets end of the straight line from Bibaibuto towards Poronai or Ichikishiri, that region of low brush and wet swamp seemed to be very long, owing to the bend in the Ikushibets which makes it for some miles run nearly parallel to the line. About noon we at length (for my own part very much fatigued) reached the Ikushibets, greatly to the surprise of the Ainos, who had all along evidently put no faith in my declaration that the Ikushibets was before us and not far off. They said that all this swampy ground was kept wet by the water of the Kawanai stream (near Nuppao-manai) which empties into it at the northern end instead of flowing into any river or lake.

These immense swamps (and the Ainos said it was the same thing on the western side of the Ishcari) were a new revelation to me ; for I had always hitherto travelled on the rivers themselves, except on the west side of the Ikushibets and near the Horumui where the swamps are comparatively small and infrequent, and the firm level ground very wide-spread. I had until we left Sapporo supposed from all I had seen and could learn by inquiry that it would be quite possible to run a straight railroad line from Horumuibuto to the hills near Poronai. So that not only has the distance in a right line turned out much greater than we had expected, but it is still further increased by the winding about that is needful to avoid the bad swamps ; and the time required for the survey is moreover lengthened by the necessity of finding out their outline and position so as to plan a road that may be as straight and short as the circumstances permit.

What I wrote from Sapporo in July about the amount of timber in the main Ishcari valley and the number of saw mills it might probably supply must also be very much modified on account of the swamp. For, instead of there being a heavy forest (as I supposed) over at least a half of the great plain of 600 square miles, it turns out that the timber land is in general only a fringe about 300 yards wide along the river banks. Even allowing for the upper parts of the streams outside of the plain it seems hardly probable that the good timber land is more than a tenth part of the 600 square miles. It is commonly the case, too, that timber grown in moist bottom lands is of inferior quality and I hear that in fact the timber of Yesso is considered poorer than that of .

Nippon. The great timber resources of Hokkaido prove therefore to be far less than we had hoped. Nevertheless several saw mills like those at Sapporo could no doubt find plenty of work to do for many years; especially if it be not thought necessary greatly to economise what timber there is.

The amount of rich land that only needs clearing of trees to become good farming land is at the same time found to be much less than I had always supposed. It is true that the open swamp lands need no clearing, and in so far there is a great saving of labor; but their present appearance is very barren. Yet I find it difficult to believe (in the absence of any experimental proof) that the swamps would not by drainage become fine farming land; for wherever a watercourse, even though small, is found its banks are for some distance firm ground covered with a very rich growth of weeds, wild hemp or ferns. The surface of the swamps is so high (a dozen feet or more) above the larger rivers and the rivers are so near, and (as we found by levelling at Bibaibuto) the surface rises so rapidly (say two feet in a thousand) in going from the top of the river bank towards the middle of the swamp; that it seems there can be no serious difficulty in drainage. It is an experiment that it would be well worth while for the Kaitakushi to make.

After eating our luncheon on the bank of the Ikushibets and resting a while, we returned to the Ishcari almost by the same way that we came. We however avoided a little of the swamp by following for a short distance near the Ikushibets the line lately cut, surveyed and staked by Mr. Takabatake for a road from Horumuibuto to Poronai, which we here came upon for the first time, and roughly surveyed for a few hundred yards. At Sapporo no one had thought it worth while to mention to me that such a line had been cut and surveyed and perhaps mapped;

but it would seem natural to have put me at the outset in possession of all such facts, which might possibly have a bearing upon our undertaking.

We reached camp again near Bibaibuto late in the afternoon well tired out ; and found our companions looking at the skin of a large yellow or brown bear that had just been brought in by some of the Ainos. On our way past Bibaidap we found the high water pouring across the narrow low neck.

It seemed pretty clear that no railroad could be built from Bibaibuto across to Poronai without at least very great expense in draining properly so wide a swamp ; expense enough at any rate to overbalance the gain of a league in distance as compared with a line from Horumuibuto, especially when the additional 13 miles (5¼ ri) of river navigation was considered. Larger vessels too could be brought to Horumuibuto than to Bibaibuto if (according to the account of the Ainos, and I have not Lt. Day's map to refer to for corroboration) if only for a short distance near Ebetsbuto the Ishcari should be deepened. But above Horumuibuto the river grows gradually shallow so that it would probably be impracticable to deepen it through the long space required, so as to enable much larger vessels to go up to the Bibai. For an exploratory line to find out by surveying the real position of Poronai with referrence to Horumuibuto, the line already so well cut out and staked by Mr. Takabatake offered great advantages.

The next morning (31st July), we went therefore in our canoes to Horumuibuto, and arrived there a little before noon. In the afternoon we began surveying Mr. Takabatake's line for exploratory purposes, and as it was already pretty well cut and staked we easily ran more than a mile before night. The weather had at length become fine, and we went on with the survey day after day; but after some half a dozen miles left Mr. Takabatake's line on the right (western) bank of the Ikushibets in order on the other side to cut across the great bends of the stream of which I had found the general shape by my rough sketching last year and year before. From time to time our camp was removed, the fourth and last time to Poronaibuto; and we finally reached the upper coal places early in the afternoon of the 7th of August, just a week after starting from Horumuibuto and found we had run in all 20.7 miles (8 ri 13 cho); each day as follows: 31st July 5,912 feet ; 1 Aug. 21,587 feet ; 2 Aug. 11,760 feet ; 3 Aug. 5,066 feet ; 4 Aug. 24,559 feet; 5 Aug. 14,147 feet; 6 Aug. 13,697 feet ; 7 Aug. 11,665 feet ; in all 108,394 feet and averaging about three miles ($1\frac{1}{8}$ ri) a day. (By Mr. Takabatake's line it is said to be $8\frac{3}{4}$ ri to the same point). We had besides lost a little time from surveying by moving once or twice with the camp and by some mapping we had done. Most of the mapping had however been put off lest some rainy weather should come; and had now at last in spite of fair weather to be attended to before the course of the railroad could be planned and staked out on the ground.

The 8th and 9th of August we were all therefore very busily occupied with office work. During the surveying two of the assistants had attended to the chaining and sometimes to the rod holding, while another followed us up with the hand level. My intention was to have him go over the railroad line after we had finished running it and level it carefully with our transit level the best instrument we have at command. The two other assistants helped with the prismatic compass and pacing by running short side lines from our main line to the edge of the swamps, or along the banks of the main streams or the edges of high ground. Especially in the

Poronai valley a good deal of such work was needed in order to make out its shape fully, so as to see whereabouts a railroad line could be run without making the grading too costly or the length needlessly great. That valley indeed turned out to be much less simple in that respect than I had supposed from the limited view furnished by walking over the path through the woods ; and it was clearly impossible to lay out a line even with much winding that should be wholly within a narrow flat along the stream ; for in many places there is no such-flat.

Hakodate; 27th Sept.—The 8th of August (Sunday) was wholly spent in office work; and so was the 9th. On the 10th the weather was cloudy and threatening; but after arranging office work for assistants I went with some coolies to the Poronai coal places and had a little digging done. For, as I was on the spot, it seemed best to do a little geological work that was desirable and to point out definitely the place where it would be best for serious coal mining to begin. We quickly found by a little careful examination that the coal beds called " L602 ca " and " L602 ba " in my letter transmitting Mr. Munroe's report on the assays of Yesso coals must be one and the same bed, and therefore probably on the average not so thick as "L602 ca " had seemed to be. The coal " L602 a " is now so opened as to be easier to measure than it was two years ago and also proves to be somewhat thinner than had been supposed, to be namely, about 4,2 feet thick. It turned out therefore that there was no object as regards thickness or other merits in making the first mine further up stream than the nearest and geologically uppermost workable coal bed, that of L576 i. The old opening at that point has its coal now covered up by a slide of earth from above ; but there will be no difficulty in reopening it, and it will be a good place to begin serious mining by a drift that will run into the hill northeasterly and soon open up a fine body of good coal above water level.

On the opposite side of the brook from that point a drift might also be run southwesterly ; but as the hill is low only little coal above water level would be opened up before reaching the other branch of the main stream within a few hundred yards. That comparatively small quantity of coal (comparatively poor also, because so near the out crop) may be neglected for the present, or until workings below water level make it desirable to have that portion of the bed drained as low as possible to save pumping. We went therefore to look for the same coal bed on the south-westerly side of the branch stream, where however it was not exposed two years ago. But now it had become well uncovered there and of good thickness and quality in the bed of the brook, and we dug hopefully into the bank alongside yet without success before the approach of night obliged us to return to camp.

Taking advantage of the fact that I was in this region I made arrangements for a visit to the Nuppaomanai coal exposures (about a ri distant) mapped by my assistants last year, but in regard to whose thickness and dip they had much doubt owing to special difficulties ; and some coolies went in advance to reopen the places for me as well as they could.

The 11th of August was rainy in the morning and very threatening almost all the rest of the day; so it was given up to office work. The 12th of August was spent in like manner at office work; since in the morning it was threatening very strongly to rain, though none fell until afternoon. As our camp was so far (one ri) from the coal, where the staking out of the rail road should begin, it was not worth while to go so far with the probability so great of having to return at once, especially as there was plenty of office work that was still pressing. The 13th of August was very rainy and was given to office work. The 14th it rained several times and I did office work, but some of the assistants ventured out to do a little surveying near by that was needed, and were caught in the rain.

On the 15th of August (Sunday) the weather at length was better, though doubtful, and with one assistant and some coolies I went to look at the Nuppaomanai coals leaving the rest of the assistants busy at office work. With some difficulty and digging we succeeded in making out satisfactorily the dip and thickness and mining merits of all the coal exposures we visited. It had seemed possible that it might be desirable after all to bring the railroad here for the first mines instead of to Poronai, but the day's work showed clearly that it was not so. The coal beds had, owing to close folding together and imperfect exposure of dip, been taken to be much thicker than they really were, and in that respect no better than the Poronai coals ; in fact the beds seemed to be the same. But at that part of Nuppaomanai the dips are very steep indeed, and that makes the beds more unfavorable for working than those at Poronai.

On the 16th of August, then, after a little office work in the morning we went to the Poronai coal to begin the running of the railroad ; and first to set the coolies at work again opening the coal. Some of them had been digging there the day before at the exposure in the brook and had clearly uncovered the edge towards the bank so that a sharply cut fault could be seen remarkably well, the reason of our difficulty the former day in finding the coal close by in the bank. It was now quite uncertain how far the continuation of the bed might be removed, but as there was a leader of coal in the fault leading northerly we dug at several points in that direction without success. With the interruption to the surveying that was caused by the coal search and digging, little more than a mere beginning was made before it was necessary to set out for camp again ; though enough was done to disclose an important error in the former year's rough surveying on which our present plan was based at that point.

On the 17th of August the coal digging and the surveying were continued, both under disadvantage from the interruptions caused by the other. The best coolies we had to spare for the coal search were so unused to that kind of work that a good part of their labor was lost even when you could stand and watch them and direct almost every stroke they made ; but when they were left to themselves it was even far worse. The coal bed was therefore not found that day, and only a few hundred feet of railroad curves in the woods over rough ground were staked out.

The 18th of August was rainy again though less so in the end than it had threatened to be, and I occupied myself all day with office work.

On the 19th of August, a few hundred feet more of railroad curves were staked out, and some digging for the coal was done, both under the same disadvantages as before. The 20th was again rainy and still more threatening, after heavy rains in the night, and we all busied ourselves with office work ; and the same was the case on the 21st. The 22nd was still very threatening, especially in the opinion of our most weather-wise man, an old Aino ; and as it was Sunday, too, we took at length a day of rest with but little office work, as what was most pressing had already been done.

On the 23rd of August we did some additional exploratory surveying, and plotted and staked out a few hundred feet of railroad curve, besides going on with the search for the coal bed. The leader of coal in the fault had proved to come from an-

other, worthless, upper bed, showing the fault to be in the
opposite direction from what had at first seemed to be the
fact; and now we at last found the bed, on the southerly
side. Much digging however seemed necessary to open it
well. On the 24th some digging was done, though at
last, in my absence, not quite enough to open the bed
well ; and the railroad line, in spite of interruption from
the digging and of very rough ground and thick woods
that required much cutting, at length fairly got away from
the immediate neighborhood of the mines and advanced
1,500 feet. Our chain men and rodmen and wood choppers
were getting a little readier at the new work required
of them. On the 25th we staked out about 2,300 feet
of the line ; on the 26th, 2,200 feet including 700 feet of
better laying out of the previous day's work. Our
exploratory compass survey had been so rapid as here
and there to require a second staking of the line, after
the more exact knowledge obtained by the first staking ;
yet a more minute exact survey in the beginning would
on the whole have required much more time. On the
27th we ran 2,900 feet of the line ; on the 28th, 3,700,
including 1,000 of correction and a little office work ; and
had now reached the neighborhood of our camp, so that
much less time and labor were daily lost in going to and
from it.

On the 29th of August (Sunday) it threatened to rain
in the morning, but I went with a couple of coolies to the
coal digging to see how it had been left ; and convinced
myself still more fully that it was the right place to begin
a mine on the bed although not yet well opened. The
bed measures 4.4 feet of good, solid coal, with 0.9 foot of
bad bony coal below that, and some 2 feet of light gray
fire clay below that. It is very possible that the fire clay
may prove valuable for making fire bricks, for which the
material is not very abundant in Japan. It could be

mined here easily in connection with the coal and give
ample height to the mine gangways. We left a large
post there (" B") with marks to show that it was the
place to open a mine. A similar stake ("A") had been
set up at " L 576 i," the place to mine the same bed on
the other branch of the brook. A branch of the railroad
line comes very near to either place.

In the afternoon of the same day, as the weather had
cleared up, we went on with the surveying and ran 1,200
feet of the railroad line. On the 30th of August it was
rainy or very threatening all day and we attended to
office work in camp. On the 31st, we ran 1,900 feet
of the line in the morning after the weather had
become a little promising; but the work was wholly
a better laying out of what had already been run once.
We returned to camp for luncheon and a little office
work, and when ready to start out again were prevented by rain that lasted the rest of the day. On the
1st of September it was rainy at times all day, and
we stayed in camp and did a little office work. On the
2nd, with fair weather, we ran 3,100 feet, of which 500
was an improvement of previous running. On the 3rd,
we ran 2,000 feet and had our camp removed from Poronaibuto to Ichikishiributo. On the 4th, we ran 5,500 feet
of the line, including 1,700 of better laying out of former
work. We had now got so far away from the intricacies
and irregularities of the Poronai valley that we could go
in a straight line most of the time, without having to stop
and turn at every hundred feet as we had had to almost
always hitherto; and the steep narrow ravines were getting fewer, so that most of the time we had nearly level
ground and were only delayed by the cutting of the still
rather thick woods.

On the 5th of September, (Sunday) it rained heavily,
and we had at last another day of rest, with but little of-

fice work to do. On the 6th, although the weather was
rather threatening at first, we ran 8,700 feet of the line,
all straight and nearly level, but with some special delays,
as in crossing the Ichikishiri brook. On returning to
camp we found that two of the assistants had come back,
who about five days before had gone to make a rough
exploratory survey for a short route up the Tonebets
valley or even further east. On the 7th, they set out
again to do additional surveying against our coming.
The same day, we ran 7,400 feet, including some curves ;
and had our camp moved a few miles further down the
Ikushibets. On the 8th, it was very rainy in the morning and very threatening the rest of the day ; and we
stayed in camp, but I had a good chance to put the transit
in very good adjustment again. On the 9th, we ran
6,350 feet, partly curves, and were driven into camp by
heavy rain in the afternoon. On the 10th, we ran 8,250
feet, partly curving ; and had our camp removed to
Tonebetsbuto. On the 11th, it was very rainy again all
day, and we stayed in camp, and busied ourselves with
office work. On the 12th, (Sunday) it was very rainy
still, most of the day ; and we stayed in camp. I wrote
a portion of the rough draft of this report. On the
13th, we ran 7,000 feet, partly curved, and ended
near our camp. On the 14th, we ran 7,000 feet,
in good part curved, crossing the Ikushibets to the
western bank; and having our camp removed a few miles
down stream. On the 15th, we ran 9,000 feet, mostly
straight and with the woods rather more open, and ended
close by our camp. On the 16th, we ran 11,300 feet, partly
curved; and had our camp removed to the place where
we ended. On t he 17th and 18th it rained heavily, and
we staid in camp; but I did a little office work, and, when
it had stopped raining on the afternoon of the 18th, did
a little rough surveying near camp. These frequent heavy

rains had raised the streams very much, and the present flood was the highest we had seen. The water rose within two or three feet of the top of the bank where we were encamped; and the lower grounds and the edges of the swamps we had to survey across were very wet indeed. On the 19th, however, we ran the remaining 6,400 feet (in good part curved) to Horumuibuto, having besides to go twice over a part of the lower end, owing to errors in previous rough surveying there, making 2,600 feet additional, or 9,000 in all. We finished about four o'clock, and, as soon as we could exchange the clothes we had been wading in for drier ones, we embarked in our canoes and went 8½ miles (3½ ri) down stream to Tsuishcari, where we encamped at night fall. For I had indirectly heard that you wished me to return to Yedo this month, and only too gladly hastened to do so.

Hakodate, 28th September.—We had, indeed, from the very beginning, as you may see by the foregoing account, worked as busily as we could, without even allowing ourselves one day of rest in the week when the weather was good or when there was office work to do; and each day's work commonly lasted from seven o'clock in the morning until six at night. We hoped by such industry to make up in some measure for the unexpected length of the line (seven ri instead of five, a difference of forty per cent.), and for the unexpected difficulties of the ground, and for the unusual raininess of the season. From the time of leaving Sapporo until the end of the survey twenty days were lost from the field work by rainy weather, 17 of which were lost after our beginning the survey from Horumuibuto, indeed after our reaching Poronai; and fully four more days were taken up by geological work ; so that only 29 days of fair weather remained for the railroad survey, including at least two or three that had to be spent in mapping. The railroad line from Yedo to Yokohama is of just about the same length and through an open, unwooded and much more level country ; you can perhaps learn on inquiry how long it took to lay it out and how many foreigners were employed on the work. We may have lost as much as a day or two in all by our inexperience ; for, not only were my assistants unused to that work, but it was the first railroad line that I had ever laid out, and many practical details that are so familiar to professed railroad engineers as to be gone through almost mechanically with scarcely a thought had to be both learned and even at first studied out by myself, with very few hints from any books I happened to have with me. The novelty of the work gave it in fact, a charm to us, and I was very glad to get a little experience in a branch of surveying that was still so new to me ; and to the young assistants "all is fish that comes to their net." I should nevertheless have hardly been willing to undertake out of complaisance a piece of work so much out of my line, if I had known beforehand that it, together with the bad weather, would have taken up so fully the whole season for geological field work ; since my services might be more valuable in a geological way and there were one or two trips that I was anxious to make this year. But the bad weather even there would have been a great drawback.

As for the desirableness of making the railroad survey at all, I was, strangely enough, not consulted at all, although I was given to understand at Yedo that my presence in Sapporo was required for such a consultation. On arriving at Sapporo I found there only comparatively inferior officials who in a couple of days showed me a telegram which said that you "consented" to the railroad survey ; at whose request I did not learn. After several days more (owing to a break in the telegraph) one of those officials verbally asked me to make the survey. I beg that in future such important orders and even less important ones may come to me direct from yourself and in writing.

The railroad line from near the stake "B" to Horumui-buto is 92,425 feet long or 17½ miles (about 7 ri) ; and there is besides a branch of 900 feet towards the stake "A." The upper end of the line for a ri or so will require a little heavy cutting and filling here and there ; but the rest is mostly very flat, with here and there, a narrow ravine or brook to cross and one bridge of a hundred feet or so on the Ikushibets. The grades are uniformly downward towards Horumuibuto and are for the most part very gentle, but the steepest at the upper end is about one in fifty or a hundred feet to the mile for a few hundred feet. To build the road and stock it with locomotives and cars would cost perhaps $700,000 ; on which the interest at ten per cent would be $70,000 a year, a tax of seventy cents a ton on a yearly product of 100,000 tons, the yield of a good colliery. If, however, the road should be worked at first with horses instead of steam, it could be built with lighter iron rails and other-wise somewhat cheaper, and the stocking of it would cost much less. The first cost might in that way be reduced perhaps to $250,000 or $300,000 ; on which the interest at ten per cent. would be only $25,000 or $30,000 or a tax of 25 or 30 cents a ton on 100,000 tons. As the grade is downward all the way from the mines, a light car could be made upon which two horses could travel all the way to Horumuibuto along with a driver and a car loaded with five tons of coal (if some portions do not prove too level), and the same horses and driver could bring back both cars empty the same day to the mines ; making a cost of perhaps 20 or 25 cents for the carriage of each ton of coal ; or, together with the interest, 50 cents in all, not counting repairs to the road and cars. For a yearly product of 100,000 tons it would therefore be much cheaper to use horses than steam. For 200,000 tons a year the cost by horse power would still be somewhat less than that by steam; for the $70,000 interest required for steam power would be divided by a larger number and become only 35 cents a ton and the mere transportation (not counting repairs to the road) would be perhaps ten cents a ton additional, or 45 cents in all, whereas the horse power would cost as before, say 23 cents a ton besides the interest say 14 cents a ton; or 37 cents a ton in all. But for 300,000 tons a year steam would become about as cheap as horses; for the yearly interest on the first cost would become only some 23 cents a ton which added to the ten cents would make but 33 cents a ton in all; whereas the horses would cost in like manner 23 cents a ton besides nine cents for interest, or 32 cents a ton in all. It must not be forgotten that these amounts (owing to repairs and such expenses) are not the full cost of carrying the coal. The calculation is a very rough one; but it is clearly enough to be seen that horse power will be cheapest for a yield of less than 300,000 tons a year, and that the larger the yield the less the cost for each ton, though diminishing more and more slightly as the number of tons becomes very large. Even with a very large yield it is likely that the whole cost including repairs and everything would be 25 cents a ton (perhaps even 30) for carrying the coal from Poronai to Horumuibuto; for the first few years with a comparatively small yield it would probably be at least 50 cents a ton. This even is very far less than what it would cost to carry the coal by log canoes or by horse power without a railroad (which some of the Sapporo officials seem to have thought of).

As compared with the Kayanoma mines there is also to

be considered the greater distance that the coal has to be carried by water towards any market ; for the Poronai coal would have to go right past Kayanoma. Sufficiently large steam vessels could no doubt carry the coal direct from Horumuibuto down the Ishcari to its mouth and by sea to Yokohama or any other port. But the time and distance between Horumuibuto and Kayanoma would cause an expense of perhaps 25 cents a ton for the coal. Adding that to the 50 cents for carriage by rail from Poronai to Horumuibuto the cost would be 75 cents a ton against the cost of carrying the Kayanoma coal by a down grade road two miles long, without steam, perhaps even without horses, if, as is probable, arrangements could be made by which the loaded cars would pull up the empty ones.

It is further to be remembered that the navigation of the Ishcari river is closed by ice during four months of the year ; so that the coal dug in the winter would have to be stacked in large heaps, exposed to the weather for several months, and consequently deteriorating in quality. If however a rail road should be built at some future time all the way from the mines to the mouth of the Ishcari, coal might be shipped throughout the year ; and, if the swamps admit of building a railroad straight across them, the distance would not be very great.

It seems therefore on the whole highly probable that in spite of the superior quality of the coal of the Ishcari valley it will be most advisable to begin active coal mining more especially at Kayanoma. Even if the little breakwater and artificial harbor at Shibui should be built for that purpose, the cost would be far less than that of the Poronai railroad. The Furushiki coal at Kayanoma, though long abandoned, is by far the best caking coal yet found in Yesso, and therefore the most suitable for the demands of iron furnaces ; and can be so mixed with other

Kayanoma coals as to make good coke out of a large proportion of them all. It is likely that more care in mining may very much reduce the amount of slate and consequently of ashes in the Kayanoma coals ; and some of the beds or parts of beds are of quite fair quality in that respect. It many even be worth while to crush and wash some of the coal, so as to purify it from the injurious stony particles before coking ; and such washing is said at one of the American collieries to cause an additional expense of only 15 cents a ton.

The Kayanoma coal field however is very much smaller than the one east of the Ishcari River, so that while working at first chiefly at Kayanoma, where the mines already opened would give a certain advantage for arriving quickly at a handsome yearly yield, mining might also be begun at Poronai, which would in spite of additional cost of carriage, help to supply the constantly increasing demand for coal, and would have the advantage of superior quality.

We ran the railroad line as if for steam power, with curves of a radius large enough for locomotives, thinking that they would be needed within a few years and that it would be cheapest on the whole to count on them from the outset ; but for coal cars and horse power alone sharper curves might in some places be allowed and consequently greater cheapness of building, as in winding along a hill side or round the edge of a swamp or a bend in a brook. For a bridle path or wagon road the best line would still further differ, as the constant down grade from the mines could in some places be abandoned for a more direct course with a slight up grade ; and in other places the cost of heavy grading would not be justified for such a road, though distance should be saved thereby. The best line for a wagon road would however not be very far from the railroad, and would be decidedly shorter than the one run by Mr. Takabatake ; and can no doubt readily be laid down on the map we shall make of our work.

On the 20th of September, while the five assistants went direct to Sapporo by the much swollen Toyohira River, my quarter master and I, with two coolies and two Aino boatmen, went down the Ishcari to visit the iron ore of Oyafuru near Ishcaributo and at Utsunai and at Bannaguro. Our baggage, servants and other attendants also went with us as far as Sapporobuto, but went thence up the Shinoro to the village of the same name, and waited for us there. On the way down the Ishcari I was struck with the increased populousness and activity of appearance of its banks from Tsuishcari to Sapporobuto since two years ago, when in the summer it seemed almost uninhabited. Now we found a great number of very large grass houses, and the men were busy every where making preparations for the salmon fishing now near at hand. Large boats and nets were numerous. Above Tsuishcari there are no such fishing houses.

We reached the upper end of Oyafuru, on the northerly bank a ri above Ishcaributo, about noon, and on landing found that some of the ore was close by. It lies on the surface of the ground, in some places it is said to be as much as a foot below it, in small patches of a few yards in diameter here and there throughout the farming settlement of half a ri in length and from the river bank back to the uncleared woods ; through a space, that is, of perhaps 150 acres. A very intelligent inhabitant, much more familiar with the whole ground than I could possibly become without a lengthy survey, estimated that the ore covered about a tenth of the whole of that surface, or say 15 acres (18,000 tsubo) ; and that is probably not underrating the extent. The thickness varies, and is at most about a foot and a half ; and the ore is so porous as to have comparatively little weight by the cubic foot, perhaps at a guess one hundred weight. If the average thickness of the 15 acres of ore should be one third of a foot, or say equal to 5 acres of one foot in thickness, the whole amount of ore would be only about 11,000 tons ; and at most it can hardly be more than 20,000 tons. So small an amount is quite too insignificant to justify building furnaces expressly for its smelting ; but if furnaces should ever be built for smelting other ore with the coal of the Ishcari velley it may very well be worth while to gather what there is of such easily accessible ore as that of Oyafuru ; if its quality from the presence of phosphorus should not be so low as to make it too bad to mix with the other ore. The quantity on clearing the woods further from the river may prove to be some what greater. The richness and quality and amount of phosphorus are probably very closely like those of the ore of Hiragishi-mura, described above, which it closely resembles ; and the origin was doubtless the same, although at Oyafuru the land is not now swampy.

From Oyafuru we worked our way slowly against the current of the swollen Ishcari one ri up stream to Utsunai, on the southerly shore ; and after some delay found that the ore place was quite near to where we had landed. It proved however to be in a swamp and at the moment under water, so that nothing could be seen even if we waded to it. It had already been described to us by an inhabitant as having extremely little ore compared with what was found at the upper end of Bannaguro, which is far less than that at Oyafuru ; so that the amount of ore is no doubt wholly insignificant. It is described as lying in small patches, say a couple of yards in diameter at most, within a space of perhaps 40 yards across. It is called 1½ ri from Ishcaributo and 2 ri from Shinoro and is close by the road between those places.

We walked then to the upper end of Bannaguro, three miles (" 1 ri and 8 cho ") from Shinoro on the same road, and were shown there two or three little patches a yard or two in diameter in a garden where lumps of the same ore from the size of a fist down were exposed on the surface. Behind the house close by was another patch four or five yards in diameter, where the ore opened at one point proved to be 0.8 foot in diameter. On the road a couple of hundred yards towards Shinoro there were lumps again up to even a foot and a half long, through a space of some 20 yards long ; and there was said to be more of it on the bank of the Ishcari, a hundred yards distant. Again about three quarters of a mile from Bannaguro towards Shinoro, where a new road has been begun, there are lumps of ore up to two feet in length, through a space about ten yards long. The appearance and probably the quality are the same in these Bannaguro ores as in the Oyafuru and Hiragishimura ones ; but the amount is still far more insignificant, so that it deserves no consideration at all.

We walked on to Shinoro and arrived there about half past four. I began to look with great admiration on the new road just mentioned, as a case where the Kaitakushi had begun to "open up" the country in an efficient way, and at a point where such facility of inter-communication must be especially desirable ; when suddenly after a few hundred yards the new road came to an end. Perhaps wood is trusted to rather too much for holding up the sides of a causeway through a swamp ; for in a few years rotting will necessarily set in. In passing through Bannaguro from end to end, I thought I could perceive a decided advance in the prosperity of its inhabitants since my visit there two years ago with General Capron. We saw this time a number of pretty well dressed men, women and children, and a number of small shops and

itinerant merchants, as if something more than a mere
subsistence had been gained from the soil. The children
looked well fed and happy.

After waiting nearly an hour in vain at Shinoro for
my horse, which had been taken to Ishcari for me, and
sent for again ; and after succeeding in getting some of
our baggage started on the few packhorses that could be
found, I started on foot again for Sapporo, $3\frac{1}{2}$ ri ; and
arrived there about eight o'clock : some of the others with
the remaining baggage did not arrive until an hour or
more later. The whole distance travelled by me in the
day was 36 miles ($14\frac{1}{2}$ ri) ; of which $22\frac{1}{2}$ miles (9 ri)
were by canoe and.$13\frac{1}{2}$ miles ($5\frac{1}{2}$ ri) were on foot.

We found that the five assistants who had been surveying towards the Sorachi had arrived the night before in
compliance with the instructions I had lately sent them.
In their survey they had reached Naie and within a ri or
two of the Sorachi coal, and had found coal at several
places on their way, in some places up to six feet in thickness ; practically proving the uninterrupted continuity of
the coal field all the way from Poronai to the Sorachi.
They had been much delayed by the remarkable wetness
of the season, and by the consequent ill health of some of
their number.

On the 21st of September towards noon we set
out for Hakodate and reached Chitose, 24 miles
($9\frac{3}{4}$ ri) that day. The next day (22nd September)
in spite of some rain we rode on to Shiraoi, 31 miles
($12\frac{1}{2}$ ri) ; and on the 23rd with good weather rode
to Tokarumui, 30 miles (12 ri). It is a tedious journey over those comparatively barren plains of pumice ;
which seem likely always to remain so, since the Tarumai
volcano will probably long continue to throw out showers
of pumice from time to time to cover up any thin soil that
may accumulate. The idea of ever making Tokarumui the

port of Sapporo seems enormously preposterous ; an admirable harbour to be sure, but separated by eighty-five miles of almost uninhabited and uninhabitable country from the metropolis, a village now of perhaps 1,500 inhabitants and probably never destined to be a populous city !

Otarunai stands a much better chance to hold such a position permanently ; but can not be the true outlet of the Ishcari valley (and consequently the future port of Sapporo) without a railroad ; and that for the nine miles next to Otarunai, would be more difficult and costly than the building and keeping in repair of an excellent artificial harbour entrance at Ishcari, which is therefore destined undoubtedly to be the great port.

We found the little ferry steamer out of order and not running to Mori, so we took a small junk, and having pushed to sea at six o'clock on the morning of the 24th, with sculling at times and with the sail at other times we succeeded in reaching Mori, 25 miles (10 ri), at six in the evening.

On the 25th of September we rode from Mori to Hakodate 29 miles (11¾ ri). The New Road proved to be in better condition than I had ever seen it, with good bridges all the way and every thing in good order, except two very small bridges, one of which is too narrow for waggons, and the other is in bad condition even for horses. With a good prospect that the road would be kept in such good repair, it would undoubtedly be worth while to put on it at least one daily line of waggons for the carriage of the goods now carried across by pack horses. The road between Sapporo and Tokarumui was also mostly in good order, except that almost every bridge was undergoing renewal, and could not at the moment be crossed. There seemed however to be none begun at Horobets.

We found no steamer for Yokohama here after all, and

so have had to wait. The 26th of September was Sunday, and we took it as a day of rest. A violent storm that had been threatening the day before finally broke with rain and high wind, and we congratulated ourselves that our exertions had brought us here in time to escape it. The 27th and to-day (28th) I have been busy report writing.

Hakodate; 29th September.—I have this season learned but little additional in regard to the Ainos, we had so few with us and there was such a want of leisure. I find, however, on inquiry of the quartermaster, who is a native of Yesso, I understand, and has seen a great deal of the Ainos and knows their language, that some statements in regard to them that have lately come under my eye in the excellent English periodical called *Nature* (Vol. IX, p. 428 : 2 April, 1874.) are not quite correct ; and some of my assistants are also able to add some information. He has never heard of the traditional origin of the Aino race represented by the picture of a woman in a cave weeping to whom a dog is bringing a red flower ; and it is possibly merely the illustration to some Japanese tradition or popular fable. It is said that the Ainos of different places have quite different stories about their origin. The Ainos do carry burdens on their back, though with the help commonly of the band across the forehead. The tattooing of the lips of the women is commonly completed at the time of marriage, though begun .often at a very early age ; but the netlike tattooing on the back of the hands or lower arms is not an invariable accompaniment. The Ainos have stringed musical instruments of three, five and six strings ; and the strings are made of the sinews of dead whales that have been cast up on the shore. The Ainos do not hunt the whale but respect it, saying that it feeds on herring, which consequently flee before it into the shallow water near land where the

Ainos take them. They say, too, that the whale eats the sardine (iwashi), a hundred at once.

The Ainos do go through with some propitiatory ceremony before eating the flesh of a bear or even of a deer they have killed ; so as not afterwards to be harmed by its spirit. They bring the bear's head home and fix it on a pole near their dwelling ; a deer's head they wrap in grass and leave behind in the woods or they set it up on a pole there. The collection of sea-weed and other fishery products for sale to the Japanese is, according to my observation, a step in their civilization above their fully wild state ; just as in some villages they have learned a little of husbandry from the Japanese. But a somewhat educated Aino at Yedo once told me that there was no word in their language for " farmer " or " husbandman." The Ainos can hardly be said to have been "driven inland by the fringe of Japanese settlements all round the coast ;" for the Aino villages are still mostly on the coast, and very few indeed live inland. They are said to have been driven gradually northward from at least the central part of Nippon ; and are now rare in the most southern part of Yesso.

The Ainos, it appears, have a great disgust at snakes, perhaps even a fear of them, and I saw it shown more than once this summer ; yet the most poisonous of the Yesso snakes appears not to be very dangerously so. It is called the "mamushi," and according to my memory closely resembles the American rattle-snake, though smaller. We found one on a morning in August, near Ichikishiributo, and the coolies killed it, and even after that in spite of our laughter an old Aino with us ran away with horror. The snake was some two feet and a half long, thick and mottled with brown. On looking for any trace there might be of a rattle at the end of the tail, there proved in fact to be a horny sharp point perhaps an eighth of an inch

long or less. The bite is said not to be fatal, though it
may cause a few day's sickness. Another day I saw at
Poronai a snake that resembles the American brown
adder, though of a brighter, more reddish color ; and was
told that its bite was poisonous, but less so than that of
the mamushi. Such resemblances between the snakes
of north-eastern America and of Yesso would seem
to correspond to like resemblances between the plants,
that have so long been remarked. The number of and
boldness, or tameness of mice, apparently of several kinds,
that we have found this season at almost every camping
ground, and while surveying in the woods, and even once
swimming in the river, have been quite astonishing. We
have also seen a number of squirrels ; and among them a
striped one much like one we have in America ; and a
black one. The insects, in this wet season have seemed
especially numerous and various. Some of these smaller
natural history objects have been preserved in carbolic
acid by the assistants, particularly by Mr. Inagaki. In
crossing Volcano Bay the other day some half a dozen
fishes perhaps 15 or 20 feet long, which we took at first
for small whales, passed close by us, rising to the surface
and diving below, some of them going under our little
junk of about nine tons. The boatmen, who beat on the
side of the junk to make a loud noise so as to frighten
them off, said they were not whales but *kamigiri* (*kami*—
above, and *kiri*—cutting) and that a stout, forward curved
fin on their back was stiff and sharp, and that with it
they would attack and kill a whale as with a sword.

Yedo, 7th October.—On the 29th of September I wrote
a part of this report ; and on the 30th was also busied a
little in connection with it, and our heaviest luggage was
put aboard the steamer. On the 1st of October we pack-
ed up the rest of our luggage and with it towards night
went aboard the *Taiheimaru,* which with favorable wea-

ther brought us in a little less than three days to Yokohama. The night of our arrival (4th Oct.) some of us came at once to Yedo; the others owing to the lateness of the hour, waited at Yokohama until the next morning.

I have to express great satisfaction at the fidelity and industry of my geological assistants, Messrs. Inagaki, Kuwada, Misawa, Takahashi, Kada, Saka, Shimada, Yamagiwa, Mayeda and Nishiyama and am most thankful for the efficient aid they have given. The five that were with me in the field were kept very hard at work almost constantly; but were always ready for any unusual exertion that was required. The other five, though not under my eye, were, I doubt not, equally faithful.

I have the honor to be,

Sir,

Your most obedient servant,

BENJ. SMITH LYMAN,
Chief Geologist and Mining Engineer
to the Kaitakushi.

GEOLOGICAL SURVEY OF HOKKAIDO.

REPORT

ON THE

MAKUMBETS COAL FIELD,

SHIDZUNAI DISTRICT, HIDAKA PROVINCE, YESSO;

ACCOMPANIED BY A GEOLOGICAL AND
TOPOGRAPHICAL MAP OF A
ROUGH SURVEY;

BY

BENJAMIN SMITH LYMAN,

CHIEF GEOLOGIST AND MINING ENGINEER TO THE
KAITAKUSHI.

TOKEI:
PUBLISHED BY THE KAITAKUSHI,
1876.

GEOLOGICAL SURVEY OF HOKKAIDO.

REPORT ON THE MAKUMBETS COAL FIELD, SHIDZUNAI
DISTRICT, HIDAKA PROVINCE, YESSO ; ACCOM-
PANIED BY A GEOLOGICAL AND TOPOGRAPHICAL
MAP OF A ROUGH SURVEY ; BY BENJAMIN SMITH
LYMAN, CHIEF GEOLOGIST AND MINING ENGINEER

The Makumbets and Sankebibai Reports of some
copies were by mistake paged separately. A few such
copies together with the Report of Progress were distributed from April to June, 1876, and can now be united
to the remainder.

.

1. *Situation.*—The Makumbets Coal Field lies in the
valley of the Makumbets, a small branch of the Shibichari River, in the Shidzunai District of Hidaka Province,
Yesso ; and is about a mile (or half a ri) above (north-east
of) the mouth of the Makumbets, and about 11 miles
($4\frac{1}{2}$ ri) north-east by river from Shibichari village on the
sea coast at the mouth of the stream of that name. That

village is nearly half way from Niikap on the west to Shidzunai on the east, about five miles (2 ri) from either; and is about 27 miles (10¾ ri) north-westerly along the coast from Uragawa. The space covered by the survey is about 2½ miles (1 ri) long, north and south, by one mile (14 cho) wide.

2. *Lay of the Land.*—The Makumbets valley near the coal mines has a general course from north to south through the middle of the survey; with a comparatively flat strip of bottom land about 200 yards wide, bordered by steep hillsides of very irregular shape. At the south-east corner of the survey is a very flat table land, or river terrace, a thousand yards long north and south by 400 yards wide; and another nearly as large but with much more broken surface around the coal mines, about the middle of the eastern half of the survey, with another patch of level land 200 yards wide on the opposite (western) side of the brook. All three of these terraces are about 200 feet higher than the brook along side, and the two upper ones are of about the same level. The land rises about 400 yards west of the brook into a ridge about 400 feet above the stream; and on the east rises to about the same height in a distance of 800 yards. The upper part of the course of the brook within the survey is very steep, and the valley very narrow without bottom land. A very narrow bridle path, scarcely more than a deer path above the mines, runs nearly throughout the limits of the survey, mainly following the principal valley.

3. *Geology.*—The beds of rock within the limits of the survey seem to have a general dip of say 45 degrees (varying from perhaps five degrees to sixty) about south 75 degrees west, but to have the general course of their strike (north 15 degrees west) varied by several subordinate gentle waves or rolls whose axes run about north 45 degrees west.

The rocks (aside from the alluvium of the bottom land) seem all to belong to what I have called in former reports the Horunni Group, the coal bearing group of other parts of Yesso ; and to be probably of late Secondary or early Tertiary age. The following is a section (from above downwards) of rocks exposed at different points and combined by means of the survey into one column showing the thickness and the relative position of the different beds.

(A rough measurement is marked ± ; a guess is marked ?; but these uncertainties do not affect the distance of a bed from the top of the column.)

Clay	3.0 ?
Coal	0.6
Clay	3.0 ?
Hidden	82.0
Clay	3.5±
Coal	0.6
Clay	5.0±
Hidden	99.0
Pebble rock	4.0 ?
Hidden	95.0
Pebbles	3.0±
Clay	0.5
Coal	0 5
Clay	0.5±
Fine sand rock	5.5±
Pebble rock	5.0±
Hidden	11.0
Olive green shales	3.0±
Coal	0.03 to 0.05
Olive green shales	3.0±
Pebble rock	3.0±
Hidden	220.0
Coal	0.2±
Bluish clay shales	2.0±
Clay 0.5 to 1.0	0.75

Coal	0.9±
Clay	41.0 ?
Yellow sand rock	59.0 ?
Shales	8.0±
Hidden	44.0
Coal	0.15
Hidden	33.0
Dark green hard close-grained sand rock	3.0
Green soft sand rock	2.0
Green sand rock	10.0
Dark green hard close-grained sand rock	5.0
Dark close-grained sand rock	6.0 ?
Hidden	29.0
Shales and sand rock containing traces of coal	7.0 ?
Hidden	4.0
Pebble rock	2.0 ?
Clay	1.0
Coal	2.8
Clay	3.0 ?
Hidden	3.0
Soft gray sandrock	11.0 ?
Hidden	67.0
Clay mixed with shales	2.0
Hidden	16.0
Shales	5.0 ?
Hidden	5.0
Bluish gray fine-grained sand rock weathering brown	5.0 ?
Green shales weathering brown	5.0
Bluish gray fine-grained sand rock weathering brown	6.0 ?
Hidden	34.0
Pebbles	2.0 ?
Fine pebble rock	2.0
Light gray sand rock	2.0

Pebble rock............................	2.0 ?
Hidden	15.0
Hard fine-grained greenish gray sand rock........................	6.0 ?
Hidden 167.0	
Shales	4.0±
Hidden.................................	21.0
Shales	4.0 ?
Hidden.................................	81.0
Grayish green shales	13.0 ?
Hidden	3.0
Gray shales...........................	9.0 ?
Hidden...................	21.0
Greenish gray shales	9.0 ?
Hidden 138.0	
Greenish gray shaly sandrock...	10.0 ?
Hidden	11.0
Dark gray shales....................	18.0 ?
Hidden	3.0
Greenish gray shales	7.0 ?
Hidden	4.0
Greenish gray shales...............	13.0 ?
Hidden	11.0
Shales	89.0 ?
Hidden	21.0
Brownish gray shales...............	4.0 ?
Blackish gray shales	0.3
Soft coal	0.5
Bony coal	0.2
Coal (in one place 2. 7)............	2.0
Blackish sandy shales...............	0.6
Rather soft coal.....................	0.65
Black coaly shales...................	3.0 ?
Hidden	9.0
Shales.................................	36.0 ?
Hidden	43.0
Coal 0.2—0.3........................	0.25
Hidden	3.0

Close-grained sandrock.............	3.0 ?
Clay	0.3±
Coal	0.5±
Sandrock	3.0 ?
Hidden........	17.0
Blackish gray shales...............	5.0 ?
Coal	2.6
Dark brown clay....................	0.3
Coal	0.2
Dark brown clay....................	0.75
Coal	0.2
Blackish gray shales...............	4.0 ?
Hidden	12.0
Clay	46.0 ?
Hidden	40.0
Shales	20.0
Coal	0.4
Shales	7.5
Hidden	33.0
COAL	**3.5**
Hidden	72.0
COAL, said to be...................	**4.0**
Hidden	20.0
Shales.....	19.0 ?
Hidden.............................	100.0
Sandy pebble rock	9.0 ?
Coal	0.2
Hidden	48.0
Sandy pebble rock	20.0 ?

2,252.50

There is great resemblance between this section
(especially in its lower half) to the upper part of the
general section on the Bibai River going to show that the
coals of both places are but parts of the same great field.

It is said that the Ainos formerly washed gold sand in
the upper part of the Shibichari valley ; sand derived
probably from rocks of another group.

4. *Coal.*—There are, then, but two coal beds that are more than three feet thick and may be called workable. The upper one of these was measured by Mr. Munroe in the bed of a small stream and found to be three feet and a half thick. There was an old mine (a drift) on the same bed, but it was inaccessible and a new and perhaps imperfect trial pit near it (by Mr. Inagaki) gave a thickness of but one foot and seven-tenths, including four-tenths of a foot of an irregular layer (a " horse ") of hard brown rock.

There was also an old drift on the lower bed that could no longer be entered ; and our only knowledge of the thickness there comes from a former miner who says apparently with some exactness that it was four feet.

There is besides an old drift on the 2. 6 foot coal bed about 170 feet above the $3\frac{1}{2}$ foot bed ; and six old drifts on the 2. 0 foot to 2. 7 foot bed that is about 295 feet above the $3\frac{1}{2}$ foot bed, five of the six being in one group very near together. All of the drifts were at the time of the survey inaccessible, so far as to the coal ; and the thickness was measured outside of the mines. It would seem, then, that most of the old mining was done upon the two uppermost beds, that appear to be too thin to work with profit, that is, decidedly less than three feet. It is said however that the workings were abandoned only because of the lack of demand for the coal when it had been carried to Shibichari on the sea shore, a point from which vessels could not well take it. The mining seems to have been done about the year 1870 by the government, and to have lasted about a year.

The outcrop of the two lower beds that seems workable run rather crookedly with a general north and south course, varying with the shape of the ground and with the change of strike, along near the eastern border of the survey ; and are in general about parallel to each other and from fifty to a hundred and fifty yards apart, the lowermost one to the east of the other. The outcrops indeed, as it turns out, lie in good part just outside the limits of the survey.

Our map enables the space covered by different portions of the beds upon it to be measured, and (reckoning the thickness also) gives the number of cubic yards (or tons) of the coal. In that way, we find within the map 66 acres (80,000 tsubo) of the upper workable bed, or (at a thickness of $3\frac{1}{2}$ feet) 454,000 tons above the lowest natural drainage level, that is, above the level of the lowest point of the out-crop of the bed (on the map), in this case about 170 feet above the sea, the lowest point upon the surveyed part of the bed from which a drift could be driven in that would drain itself without the help of pumps. A good portion of the space from that level on the bed upwards to the outcrop happens to lie outside the limits of the survey. The space on the map from the lowest drainage level on the same bed down to a level five hundred feet lower is 247 acres (299,000 tsubo) giving 1,581,000 tons ; making in all from that lower level up to the outcrop 313 acres (379,000 tsubo) or 2,085,000 tons.

The lower workable bed in like manner has within the map limits 67 acres (81,000 tsubo), or (at a thickness of four feet), 528,000 tons above the lowest drainage level (about 150 feet above the sea) ; and 216 acres (261,000 tsubo), or 1,637,000 tons between that level and 500 feet lower ; making in all above the lower level 283 acres (342,000 tsubo), or 2,165,000 tons.

Both beds, then, within those limits would have 982,000 tons above the lowest natural drainage level ; 3,218,000 tons between that level and one 500 feet lower ; or 4,200,000 tons in all. As one bed lies over the other through the greater part of those spaces, the extent of land that covers the portions of both that are above the drainage level is but 95 acres ; between that and the 500 foot lower level 275 acres (332,000 tsubo) ; and the whole of both portions of both beds 342 acres (414,000 tsubo). Each bed probably has for the next 500 feet of depth quite as much more coal as in the 500 feet just measured ; so that the two beds together would have within the moderate mining depth of a thousand feet below natural drainage somewhere about 7,400,000 tons ; a quantity that should, however, be increased by the portions of considerable size near the outcrop that lie just outside the map and cannot therefore be measured now.*

The quality of the coal does not seem to be remarkably good ; but no assay of it has yet been made.

5. *Shipments.*—There is no road from the mines to Shibichari on the sea shore (about 11 miles, or 4½ ri) except a bridle path ; but that is mostly very level, and a waggon road or a tramroad could be built without difficulty. It is said that the road from Shibichari to Uragawa (27 miles, or 10¾ ri) is every where along the seashore and level, so that a waggon road or tramroad could easily be built. At Uragawa is a somewhat sheltered roadstead or harbour, and there is no other so near that is anything like so good. In the other direction from Shibichari the

---

* It has been discovered within the last few days that Mr. Munroe and his interpreter measured a coal exposure very near the old mine upon the lower workable coal bed, an exposure that seems now quite possibly to be on the same bed. The thickness was only about a foot and a half of coal, separated into two halves by about a foot and a half of black clay, making about three feet in all ; a thickness that would, with allowance for the not unusual popular exaggeration in such matters, correspond perhaps well enough with the "reported thickness of four feet" for that bed. If the exposure, then, was an undisturbed one and gave the full thickness of the "lower bed," the whole amount of that bed as given above must be reckoned as no part of the workable coal of the tract.

Besides the amount of coal given in the text there are probably 810 acres or 5,300,000 tons of the upper bed within the tract between 500 and at most say 3000 feet below water level ; making the whole amount of that bed alone about 1,123 acres or 7,400,000 tons.—12 Feb., 1876.—B. S. L.

road (the usual bridle path) is said to be along the sea
shore and level, except here and there a low hill of 20 or
30 feet, as far as Yubuts (37½ miles or 15 ri.) Near
Yubuts may some day pass a railroad connecting the
Ishcari Valley and its coal mines with the excellent harbor
of Mororan ; and the distance from Yubuts to Mororan is
52 miles (20⅔ ri), and all level except two not very
impracticable hills near the port.   As there is said to be
coal at two or three places on the streams between Shibichari and Yubuts, it is not at all impossible that for the sake
of coal alone, apart from other reasons, it may be at some
future time found worth while to build a railroad along
the coast ; and in that case the whole expense would not
come upon the Makumbets field alone. There has also
been some talk of improving the Uragawa harbor by a
breakwater.

6. *Map.*—The survey was begun (in Sept. 1873) by
Mr. H. S. Munroe, Assistant Geologist, who spent about
a week there, visited most of the mines and some of the
outside exposures of rock and coal, determined the
true north by observation of the pole star, arranged
the general plan of the survey, and left written instructions in regard to it with Messrs. T. Inagaki, J.
Takahashi, T. Saito, J. Shimada and S. Mayeda, Assistant Geologists. They completed the survey in the
manner that we adopted at the outset for all our
large surveys in Yesso, running (with prismatic compass and pacing) rectangular and parallel lines, and
levelling (with hand level) stakes 200 feet apart, and
running some streams in addition. They also ran a main
line lengthwise of the survey and a cross line on either
side with transit and stadia as a check to the other lines.
Each assistant plotted the portion of the field that he had
surveyed and drew its contour lines ; but the whole was
united into one map by Mr. Inagaki, aided by Mr. Saito.

Mr. Shimada drew, under my guidance the six cross sections of rock structure and reduced them to columnar sections of the rock beds, afterwards combining them into one. The mapping of the coal beds was done by myself.

The map (including sections and all, $2\frac{2}{3}$ feet by $1\frac{2}{3}$) is on a scale of $\frac{1}{5000}$ of nature, and shows the shape of the ground by contour lines ten feet apart in level. At starting, their height above the sea was determined by aneroid readings. The outcrop of each of the two workable coal beds is laid down on the map, as well as the course that a drift would take if driven level into the hill from the lowest point on the outcrop. Of the upper ($3\frac{1}{2}$ foot) bed similar level lines are drawn for every hundred feet of level from the outcrop down to 500 feet below the lowest natural drainage level. The details of the map and of the survey are much the same as those already described for the maps of the Yamukushinai, the Washinoki and the Idzumisawa Oil Lands already in print and need not be repeated here.

I have the honor, to be,

Sir,

Your most obedient servant,

BENJ. SMITH LYMAN,

Chief Geologist and Mining Engineer
to the Kaitakushi.

Sapporo, 17th July, 1875.

GEOLOGICAL SURVEY OF HOKKAIDO.

REPORT

ON THE

SANKEBIBAI

AND

NAIE COAL SURVEY OF 1875;

ACCOMPANIED BY A GEOLOGICAL AND
TOPOGRAPHICAL MAP;

BY

BENJAMIN SMITH LYMAN,

CHIEF GEOLOGIST AND MINING ENGINEER TO THE
KAITAKUSHI.

TOKEI:
PUBLISHED BY THE KAITAKUSHI,
1876.

REPORT ON THE SANKEBIBAI AND NAIE COAL SURVEY
OF 1875 ; ACCOMPANIED BY A GEOLOGICAL AND
TOPOGRAPHICAL MAP ; BY BENJAMIN SMITH LY-
MAN, CHIEF GEOLOGIST AND MINING ENGINEER.

1.—SITUATION.
2.—LAY OF THE LAND.
3.—GEOLOGY.
4.—COAL.
5.—SHIPMENT.
6.—MAP.

HIS EXCELLENCY K. KURODA,
Kaitakuchokuwan.

SIR,—I have the honor to make you the following
report on the Sankebibai and Naie (or Naiye) coal survey
of last year.

1. *Situation.*—The survey covers a space of about
2,300 feet in width, one half of it on either side of a
straight line running about six miles N 50° E (magnetic)
from the Bibai coal survey of the year before last towards
the Sorachi river ; and covers therefore 2,6 square miles.
The line was run in order, if possible, to trace roughly
at least the limits of the great Ishcari coal field. At
every thousand feet of the main line side lines of a
thousand feet in length were run right and left, and the
course of some of the principal streams was also run.

The Epanaomap valley was also visited by Mr. Kuwada, assistant Geologist at a point about half a mile north west of the northeastern end of the survey ; and materials were obtained there for a topographical and geological sketch map of a thousand feet by five hundred.

2. *Lay of the Land.*—The survey lies near the northwestern edge of the highlands east of the Ishcari river ; and is traversed more or less nearly at right angles by the upper waters of the Sankebibai (sometimes called by mistake Ponbibai, or Little (branch of the) Bibai) on the southwest, the Chaushinai in the middle and the Naie towards the northeast, all flowing towards the Ishcari. The Sankebibai valley is, in the survey, from 250 to 300 feet above the sea and the two other valleys about 200 feet higher. The hills southwest of the Chaushinai rise to a height of some 300 feet above the two main valleys ; but towards the Naie rise 600 feet above that stream, and still further northeasterly form a large, high mass of which the highest points are even 1300 or 1400 feet above the sea. Along the Sankebibai there is a little flat land a couple of hundred yards wide ; but the Naie River, though a wider stream, has a very narrow steep-sided valley, with, in one place, a very small flat terrace about 150 feet above the stream. The hillsides almost everywhere else are quite steep. The land is everywhere covered with forest and with thick bamboo grass (Arundinaria) very tall and stout, a great hindrance to surveying.

3.—*Geology.*—The rock beds of the survey seem to lie in the form of three saddles with basins between and with axes running about due north (magnetic). One of the basins crosses the Sankebibai, and has on the westerly side dips of about 25 degrees, and on the easterly side about 60 degrees. It is followed to the east by a sharp saddle with dips of some 45 degrees eastward, forming apparent-

ly a wide basin as far as to the Chaushinai, where the
dip seems to be vertical or extremely steep towards the
west. Then a broad saddle seems to fill the space as far
as to the Naie, and to have along that stream near the
bottom of the adjoining basin easterly dips of some 15
degrees followed quickly on the eastern side of the basin
by westerly dips of about 40 °. From the Naie to the
north-eastern end of the survey no exposures of rock
were found.

The westerly dips, then are generally steeper than the
easterly ones. The basins and saddles all seem to plunge
decidedly southward. The direction of the axes is one of
the two principal axial directions that can be discerned in
the much contorted folds of the rock beds of Nuppaomanai
and that can be traced in the beds of the still nearer Bibai
coal field. But in those fields the northeasterly axial direction is more striking; and at Poronai is almost the only
one to be seen, leaving the shape of the rock folds much
more simple. It is quite possible that the northeasterly
direction might also be found to exist in the Sankebibai
and Naie field if the geology were more clearly made out
by the discovery of more numerous rock exposures and
dips.

The rocks of the present survey belong, it appears,
wholly to the lower part of the section of the Bibai coal
field and to a still lower depth ; and the following section
(downwards) has been made out from a combination of
all the observations in the field by means of the topographical and geological map :

	Feet.
Light brown shales and coaly matter	1.50
Coal (U 743 n) same as (E 344 d) of the Bibai survey	0.75
Light brown shales and coaly matter, exposed, say	3.00
Unknown	21.00

Light brown shales, exposed, say...	...	...	3.00
Shales and coaly matter (U 743 m)	...	...	1.00
Light brown shales, exposed, say ...	...	...	4.00
Unknown	...	...	198.00
Coal mixed with shales (K 1246 f) same as (B 889 y), Bibai survey ...	...	...	0.50
Bluish gray rough sand rock ...	...	...	20.00
Unknown	...	...	1.50
Gray sand rock, weathered brown (U 863 b)	...		20.00
Unknown	...	...	36.00
Gray sandrock, weathered brown...	...	...	4.50
Gray shales (U 864 a), exposed, say	...	...	6.00
Unknown	...	...	18.75
Hard brownish shales	...	...	30.00
Bad coal (N 779 f)	...	...	0.50
Yellowish gray clay, about... ...	...	...	8.00
Unknown	...	...	13.00
Hard brownish shales (N 780 h), perhaps	...		7.00
Unknown	...	...	58.00
Hard brownish shales	...	...	11.50
Brownish shales	...	...	3.50
Coal	...	...	1.00
Gray clay	...	...	3.30
Good coal (N 779 b); same as (E 421 h) Bibai Survey...	...	...	4.50
Unknown	...	...	60.00
Bluish gray sand rock and limestone balls (K 1246 b) exposed, say, ...	...	...	3.00
Gray sand rock weathered brown, say	...	...	7.00
Gray shales (U 865 b), say, ...	...	...	6.00
Blackish gray shales with fossil leaves (K 1246 c) exposed, say	...	...	3.00
Unknown	...	...	147.00
Light brownish shales (N 765 a) ...	...	...	7.00
Unknown	...	...	38.00
Gray sand rock (U 866 f), perhaps	...	...	9.00
Unknown	...	...	5.00

Gray sandrock weathered brown	10.00
Gray shales	4.02
Gray sandrock, weathered brownish, with fossil shells (U 866 ca)	1.50
Dark gray shales	1.00
Gray shales, weathered brown, exposed, say, ...	2.00
Unknown	5.00
Hard gray clay, exposed, say, ...	3.00
Coal and clay alternate (Q 754 qa) same as (E 423 r) Bibai survey, ...	1.50
Hard gray clay, exposed, say, ...	2.00
Unknown	12.00
Gray sand rock (U 866 ea), exposed, say, ...	8.00
Unknown	113.00
Gray sandy shales, weathered brown, exposed, say,	4.00
Coal shales (U 867 ia)	2.50
Unknown	15.00
Gray sandrock, weathered brown, (U 867 i) exposed, say	9.00
Coarse grained gray sand rock	12.00
Coal (N 739 da)	0.80
Unknown	93.00
Coarse grained gray sand rock (N 758 an) ...	7.00
Unknown	19.00
Gray sand rock, weathered brown...	7.00
Gray sand rock, with fossil shells and small silicious pebbles	1.00
Gray sand rock, weathered brown (U 867 l) ...	5.00
Unknown	430.00
Very fine black weathered sand rock (Q 747 cd), say	6.00
Unknown	10.00
Clay shales (Q 747 ce) say...	4.00
	1553.02

The little sketch map of part of the Epanaomap valley gives in like manner the following rough section (downwards):

	Feet.
Hard gray clay, exposed perhaps...	0.50
Coal (Q 784 n) ; same as (B 926 f)	2.00
Hard gray clay, weathered brown	2.30
Good coal	0.50
Hard gray clay, exposed, perhaps...	0.50
Unknown	74.00
Very fine hard sand rock	5.00
Very fine and very hard sandy clay	2.20
Coal with amber (Q 783 l)...	0.50
Hard dark clay, exposed, perhaps...	0.50
Unknown	21.00
Hard dark gray clay, exposed, perhaps	3.00
Coal (Q 782 k), like (Q 781 h) ; same as (K 1246 f) and (B 889 y)	1.90
Hard dark clay, exposed perhaps	2.00
Unknown	40.00
Hard very fine sand rock, exposed perhaps ...	1.00
Unknown	133.00
Hard dark gray clay	7.00
Coal (Q 781 h) ; same as (U 741 b) (and (E 420 d) ?)	1.90
Hard dark clay	3.00
Unknown	74.00
Hard dark clay, exposed perhaps	0.50
Bony coal	1.80
Coal	0.45
Coal slate	0.60
Coal...	0.15
Bony coal	1.00
Coal	0.20
Hard clay	0.35
Coal	0.45
Coal and bony coal (Q 780 g) ; same as (N 779 b) and (E 421 h)...	2.20

Hard dark clay	...	...	...	...	...	0.60
Brown sandy clay, exposed perhaps			...	...		1.00
Unknown ...	...	...	...	...	...	132.00

Greenish gray, hard, very fine sandrock, weathered brown (Q 779 da), exposed perhaps ... 1.00

Unknown 126.00

Greenish gray, hard, very fine sandrock, (Q 778 aa),
exposed perhaps 1.00

626.10

The thickness of the rough section seems to be exaggerated about twelve per cent ; but otherwise agrees well with the corresponding part of the sections of the main survey and of the Bibai survey.

4. *Coal.*—The only mineral of economical importance within the limits of the survey is the coal, and of that only one bed seems to be workable, the bed marked in the section as (N 779 b) ; a bed that in the neighboring Bibai survey is hardly good enough in quality and thickness to be workable. About 175 feet below it is the lower workable coal of the Bibai survey averaging at four places 3.22 feet in thickness ; but it was not found anywhere exposed in the present survey. It may perhaps have thinned out in this direction so as to become worthless ; while the bed of (N 779 b), the Sankebibai bed, as it may be called, by its improvement takes its place in economical value. It is safest not to count upon more than one good bed out of the two.

The outcrop of the Sankebibai coal bed crosses the river of that name twice (in shape of the letter U) near the top of the saddle or anticlinal there, and runs on the one hand northerly nearly to the edge of the survey, and, sweeping round to the west, crosses the river again and continues southwesterly to the survey limits ; while on the other hand it runs northerly to the edge of the map.

On the Chaushinai it probably crosses the survey in a southeasterly direction, chiefly to the east of the river ; but soon reenters the field again from the south and with a general northerly course crosses the dividing ridge and bending northeasterly reaches the Naie, and then with a long bend northerly passes round the rock basin there and with a southerly course runs to the border of the map. There is however no exposure of the coal yet known to the east of the Sankebibai valley ; so that the merits of the bed throughout a great part of the survey are yet unknown ; and the position of the outcrop towards the Naie, owing to the comparative rareness of any rock exposures observed there, cannot be laid down with very great certainty.

The bed is exposed not only at (N 779 b) and (Q 780 g) but also partially at (Q 757 xb) showing 3.60 feet of coal (at the top, without roof), bone coal and soft clay ; making, in all, three exposures with an average of perhaps 5.75 feet of coal, including some bony coal. As some of the bony coal may have to be altogether rejected it will perhaps not be safe to count upon an average thickness of more than four feet of good coal. No assay of the coal has yet been made,

The map enables the space filled by the coal bed to be measured, and thereby, taking the steepness of the dip into account, and reckoning a cubic yard to the ton, gives the weight. The extent of the bed above its lowest natural drainage level (about 240 feet above the sea) within the limits of the map appears then to be 170 acres, giving at a thickness of four feet 1,700,000 tons. Between that drainage level and a level five hundred feet lower there would in like manner be within the survey 220 acres or 3,500,000 tons. But going down to the bottom of the basins, at most only about 2,000 feet below drainage level, there would probably be 310 acres more, or 2,500,000 tons ; making the whole amount of workable coal 700 ✓ acres or 6,000,000 tons.

5, *Shipment.*—There are at present no roads nor navigable streams leading from the coal field towards any market ; but the Sankebibai, Chaushinai and Naie valleys enable roads to be made without serious difficulty with a continuous and gentle down grade towards the Ishcari river about five miles (2 ri) distant. The lower half of that distance is probably a flat plain, and along the banks of the Chaushinai and Naie rivers would probably be firm ground ; and the road down the Sankebibai, owing to the fact that the river ends in a swamp, would have to take the course of the lower part of the Chaushinai river, which would, however, not materially increase the distance to the Ishcari. The Ishcari from the mouths of the Naie and Chaushinai has probably a depth great enough for steamers or barges of at least five feet draught even in a very moderate height of water ; and probably at least a couple of feet more through the greater part of the year. Some day, moreover, there will probably be a railroad through the Ishcari valley as far as to the thick beds of very fine coal on the Sorachi river, about ten miles (four ri) beyond the Naie, and such a road would perhaps pass near the foot of the hills about a league only from the present survey.

6. *Map.*—The surveying and the geological observations were done in the summer of 1875 by five of the assistant geologists Messrs. T. Kuwada (in charge of the party), S. Misawa, T. Kada, S. Mayeda and S. Nishiyama under such instructions as I could give without personally visiting the field. The main line was run with a transit and chain ; the side lines with prismatic compass and pacing ; and the stations (commonly 200 feet apart) were levelled with hand levels. The topographical mapping of each line was done by its surveyor, and the

whole united into one map by Mr. Kada. The geological mapping and sections were done by myself with the aid of Messrs. Kuwada and Kada. The lettering has been written by Mr. Kuwada with the aid of Mr. J. Adachi. The whole work of the final map has been done under my own eye.

The map shows the shape of the ground by contour lines ten feet apart in level down to 500 feet below water level, with a double line for the lowest water level. The place of probable outcrop is also very clearly marked by short cross hatched lines. The staked stations are marked as in our other maps with separate numbers or letters, so that any given point of the map may easily be found on the ground.

The dips and strikes of the rocks are marked by a new method that I devised early in the winter, in order to give the direction of the strike and dip and the amount of the dip more compactly, especially for convenience in maps of a small scale and with numerous adjacent dips, like our sketch geological map of Yesso now in the engraver's hands. Instead of the usual arrow for the dip, an arrow of one barb is used, of which the shaft shows the direction of the strike while the barb (one half the length of the shaft) shows on which side the dip is (at right angles of course to the strike); and the angle of the barb with the shaft shows the amount of the dip. A break in either shaft or barb shows that the measurement of direction or amount is only a rough one; two breaks shows that it is a guess. A slightly, doubly curved shaft or barb shows the general direction or amount in case of their varying; a nearly completely circular curve in the lines shows a rock exposure with confusion or uncertainty of dip and strike. A saddle or anticlinal is shown by a doubly barbed arrow (somewhat like a letter tee), each barb showing the dip on its own side of the saddle ; a basin or synclinal by a mark like a letter wye (that is with the two barbs of the arrow drawn in the opposite direction to what is usual). A horizontal rock is marked by the shaft of the arrow crossed by a very short line or small arrow head (as it were the free end of the barb) between the middle of the shaft and its point. The rock exposure is always at the point of the barbed arrow. The method seems in compactness and simplicity to have several advantages for general use over that hitherto most commonly used for marking dips.

I have the honor to be,

Sir,

Your most obedient Servant,

BENJ. SMITH LYMAN,
Chief Geologist and Mining Engineer.

Shiba ; 12 April, 1876.

GEOLOGICAL SURVEY OF HOKKAIDO.

A REPORT

ON THE

BIBAI

COAL SURVEY OF 1874;

ACCOMPANIED BY A GEOLOGICAL AND
TOPOGRAPHICAL MAP OF A ROUGH
SURVEY ;

BY

BENJAMIN SMITH LYMAN,

CHIEF GEOLOGIST AND MINING ENGINEER.

TOKEI:
PUBLISHED BY THE KAITAKUSHI,
1876,

GEOLOGICAL SURVEY OF HOKKAIDO.

A REPORT ON THE BIBAI COAL SURVEY OF 1864; AC-
COMPANIED BY A GEOLOGICAL TOPOGRAPHICAL MAP
OF A ROUGH SURVEY; BY BENJAMIN SMITH LYMAN
CHIEF GEOLOGIST AND MINING ENGINEER.

1.—SITUATION.
2.—LAY OF THE LAND.
3.—GEOLOGY.
4.—COAL.
5.—SHIPMENT.
6.—MAP.

HIS EXCELLENCY, K. KURODA,

Kaitakuchohuwan.

SIR:

I have the honor to make the following report on the
Bibai Coal Survey of year before last.

1. *Situation.*—The survey lies in the upper part of the
Bibai valley and reaches from the Nuppaomanai survey to
the Sankebibai and Naie survey, covering a space of
riregular shape about four miles and a half (1 ri 29 chô)
long from north to south and about two miles and a half
(one ri) wide from east to west, just north of the middle;
and the whole extent of the tract amounts to 3,055 acres
(3,770,000 tsubo), or 4.8 square miles.

2. *Lay of the Land.*—The Bibai River crosses the
survey at its widest part in a general westerly direction,

and the southern part of the tract is filled up by the valleys of the Ôtakisawa on the west and of its branch Nagatakisawa on the east, small streams that flow in general nearly parallel northwards towards the Bibai. Near the north edge of the tract a still smaller stream flows eastward to join the Bibai outside the survey.

The valley of the Bibai is comparatively flat in the lower third of its course through the tract, forming a triangular space of gentle slopes some three quarters of a mile wide and about 250 feet above the sea. But all the rest of the southern part of the survey is filled by high hills which leave no level ground near the streams and grow gradually higher from something like 1,000 feet above the sea near the Bibai to about 2,000 feet at the dividing ridge between the valley of that river and the waters of the Nuppaomanai at the very southern end of the survey. On the north of the Bibai the land rises generally with more gentle slopes to a height of about 400 feet above the sea at the dividing ridge which separates us from the Sankebibai valley. The land is everywhere covered with forest and with thick bamboo grass (*arundinaria*).

3. *Geology.*—The rocks of the Ôtakisawa and Nagatakisawa valleys (if you consider the shape of any single bed) lie in the form of a large saddle (or anticlinal) or of an egg-shaped boss with the broader end towards the north-east, where it is touched by a similar saddle (or anticlinal) which prolongs the fold north-eastward. On the north is a pretty deep basin in the beds, rising northerly in a saddle form again on the edge of the Sankebibai survey. The axes of the saddles and basins may be seen to be mainly in a north-easterly and south-westerly direction but small depressions or accessory basins may be detected upon the sides of the larger ones and with axes running in a northerly and southerly direction. The same

two axial directions can be seen also in the Nuppaomanai
and in the Poronai survey still further southwest. At
Poronai the northerly and southerly axial direction is
almost completely overshadowed by the other one ; but in
the Sankebibai and Naie survey becomes so much more
prominent as to be the only one yet discovered.

The rocks of the Bibai survey belong to the Horumui
group (probably early tertiary or late secondary) ; and
appear to lie wholly below those of Poronai and to correspond in the upper part to the lower rocks of the
Nuppaomanai section and in the lower part to those of
the Sankebibai and Naie survey. The following is a
section from above downwards of the rocks of the Bibai
valley as made out by combining all the field observations
through the geological and topographical map :

	Feet.
Bluish gray shale, exposed perhaps......................	8.00
Coal slate (B 952 h).......................................	3.00
Hidden.......	13.00
Bluish gray shale, exposed perhaps	6.00
Bituminous clay................	1.00
Good *coal* 0.20	
Coal slate........................... ....2.00	
Good *coal*...................................0.10—	2.30
Coal slate..	0.80
Bluish gray shale, perhaps......·........	18.00
Coal slate, perhaps.....................................	3.00
Hard greenish gray sand rock...........................	5.50
Bluish gray shale..	13.00
Bituminous clay	1.00
Good *coal* (B 952 i).................................	0.70
Coal slate......................................	0.80
Bluish gray shale, perhaps.............................	6.00
Sand rock, exposed perhaps...........................	10.00
Hidden	11.00

	Feet.
Bluish gray shale, exposed perhaps	6.00
Bad *coal* (Y 721 c)...	2.00
Bluish gray shaly sand rock, exposed perhaps.......	6.00
Hidden	227.00
Good *coal*..0.25	
Sand rock 0.15	
Bad *coal* (B 869 bb)..............................0.35—	0.75
Coal slate.................	0.35
Hidden	28.00
Hard brown sand rock.....................................	2.00
Rather hard brown sandy clay..........................	2.00
Reddish brown bad *coal*............................0.30	
Soft gray clay...0.10	
Coal..:.......0.10	
Soft gray clay...0.30	
Coal 0.10	
Reddish brown soft clay.............................0.20	
Hard gray clay 1.10	
Soft whitish brown clay.............................0.30	
Hard dark gray clay........... ....0.50	
Coal 0.10	
Soft dark gray clay..................................0.30	
Coal 0.60	
Soft dark gray clay..................................0.20	
Bad *coal*...0.45	
Soft light gray clay..................................0.10	
Coal 0.15	
Soft light gray clay..................................0.05	
Bad coal (Q 532 a)...................................0.80—	5.75
Hard reddish brown clay, exposed perhaps............	3.00
Hidden	15.00
Reddish gray coarse grit (Y 629 jb), exposed	
perhaps	3.00
Hidden	34.00

	Feet.
Gray shale (B 952 f), exposed perhaps	14.00
Hidden	155.00
Bluish gray hard sand rock, containing fossils, exposed perhaps	5.00
Coal and coal slate, (Y 721 b)	8.00
Bluish gray hard sand rock, containing fossils, exposed perhaps	7.00
Hidden	42.00
Brownish gray shale, exposed about	20.00
Coal, an irregular streak	—
Fine grained gray sand rock, from 0 to 5 feet, say	3.00
Reddish gray soft sand rock, about	0.30
Greenish gray shale, about	7.00
Fine grained gray sand rock, about	1.00
Sand rock with fossil shells, about	2.00
Hard brown shale	6.00
Sand rock with fossil shells, about	13.00
Reddish brown shale, about	12.00
Coal with layers of grayish brown clay	6.00
Clay	10.00
Coal (B 884 c), about 3.00	
Brownish gray clay 1.00	
Good *coal* ....0.70—	4.70
Hard brownish gray shale, about	10.00
Bluish gray shale, about	2.00
Bluish gray soft sand rock about	5.00
Hidden	48.00
Bluish gray soft clayey sand rock, about	1.70
Bluish gray sand rock, (B 884 f), about	1.70
Hidden	14.00
Bluish gray sand rock, exposed perhaps	2.00
Coal (B 884 g), about	0.40
Hidden	22.00
Soft clay, exposed perhaps	1.50

	Feet.
Bad *coal* (U 533 j)..	2.50
Hard clay, exposed perhaps............................	1.50
Hidden	5.00
Soft clay, exposed perhaps............................	2.00
Good *coal* (U 533 k).....................................	1.20
Greenish gray shale, exposed perhaps.............	2.00
Hidden	10.00
Greenish gray shale, exposed perhaps.............	2.00
Coal (U 533 l)...	0.10
Greenish shale, exposed perhaps.....................	2.00
Hidden	33.00
Coal slate, about...	0.50
Coal (B 884 i), about....................................	0.30
Coal slate, about...	0.50
Hidden	17.50
Bad *coal*, (U 533 h)..............................2.50	
Gray clay...0.50	
Coal..0.20—	3.20
Brown clay..	0.30
Greenish gray shales.	1.50
Hidden	62.00
Bluish gray clay shales with fossils, exposed perhaps.	3.00
Coal, pretty good.................................3.16	
Coal slate...1.16	
Coal, pretty good (B 884 k), Upper Bed...3.16—	7.48
Dark gray clayey sand rock, exposed perhaps.......	2.00
Hidden	15.00
Greenish gray shales, exposed perhaps.............	1.00
Bad *coal* (U 532 f)	0.70
Greenish gray shales, exposed perhaps.............	1.00
Hidden	32.00
Greenish gray shales, exposed perhaps	1.50
Good *coal* (U 532 d)	0.80
Greenish gray shales, exposed perhaps	1.50

	Feet.
Hidden	15.00
Soft clay, exposed perhaps	1.50
Good *coal* 0.50	
Coal slate 0.80	
Bad *coal* (U 532 c)..........1.10—	2.40
Coal slate, exposed perhaps	1.00
Soft clay, exposed perhaps	1.00
Good coal (U 532 b)	0.75
Clay shales, exposed perhaps	2.00
Hidden	20.00
Sand rock, exposed perhaps	2.00
Light brownish gray shales, about	5 00
Coal, bad and soft, (B 884 ma)	7.70
Light brownish gray soft clay	1.70
Greenish gray hard sand rock containing much *coaly* matter	2.50
Light gray hard shales, gradually becoming sandy, exposed perhaps	3.00
Hidden	66.00
Greenish gray sand rock, exposed perhaps	7.00
Greenish gray sand rock with fossil shells (B 885 o), about	4.00
Greenish gray sand rock, exposed perhaps	9.00
Hidden	20.00
Hard greenish gray sand rock	2.00
Soft greenish gray sand rock	5.00
Gray pebbles with dark gray and blackish quartzite pebbles (B 931 j)	4.00
Greenish gray sandy shales	12.00
Fine greenish gray sand rock, perhaps	10.00
Hidden	30.00
Greenish gray shales, about	3.00
Limestone (B 885 p)	1.00
Greenish gray sand rock, perhaps	10.00

	Feet.
Many thin layers of shales of different colors......	2.00
Greenish gray shales, perhaps...........................	5.00
Hidden	40.00
Gray shales, exposed perhaps...........................	6.00
Rotten bony *coal*................................1.00	
Soft bluish clay.....................................0.15	
Firm good *coal*, (U 564 k) ....2.30—	3.45
Gray shales, exposed perhaps...........................	5.00
Hidden	9.00
Gray pebble rock with dark gray and blackish quartzite pebbles...................................	2.00
Greenish gray shale....................................	6.00
Coal slate (B 845 r).....................................	0.40
Greenish gray sand rock, about........................	6.60
Hidden	5.00
Greenish gray sand rock, perhaps.....................	10.00
Hidden	70.00
Soft clay, exposed perhaps.............................	2.00
Bony *coal*, (U 566 d).....................................	1.10
Gray shale, exposed perhaps.....................	4.00
Hidden	2.00
Firm good *coal*, (U 564 h).............................	1.10
Gray shale, exposed perhaps...........................	10.00
Hidden 185.00	
Sandy gray shales, exposed perhaps...................	5.00
Soft *coal*, (U 563 ea)	0.30
Sandy gray shales, exposed perhaps	5.00
Hidden 106 00	
Greenish gray sand rock.................................	5.00
Gray pebble rock with dark gray and blackish quartzite pebbles, (B 887 ea)..................	3.00
Greenish gray sand rock.................................	4.00
Hidden	15.00
Greenish gray sand rock (B 930 e)....................	20.00

	Feet.
Hidden	65.00
Greenish gray sand rock (B 888 ha), perhaps......	10.00
Hidden	14.00
Soft clay, exposed perhaps	2.00
Coal, good but soft, (U 572 cb)...........................	0.80
Grayish brown clay shale, exposed perhaps.........	2.00
Hidden	85.00
Coal with many layers of reddish or brownish clay	
and shale (Y 669 nb), about...................	1.70
Hidden	10.00
Lime rock (B 888 j).......................................	0.50
Greenish gray sand rock.................................	1.00
Lime rock...	0.50
Hidden	38.00
Grayish white sand rock (B 888 la)...................	1.50
Hidden....................	5.00
Greenish gray shale (B 888 lb), perhaps.............	3.00
Hidden, perhaps...	7.00
Greenish gray shale (B 888 m), exposed per-	
haps.......................................20.00	
Greenish gray shale, exposed perhaps....... 5.00—	25.00
Bad *coal* much mixed with earth (B 888 n).........	0.10
Greenish gray shale.......................................	1.00
Hidden	25.00
Lime rock.. ...	1.00
Greenish gray sand rock with fossil leaves	2.00
Coal slate	0.70
Hard greenish gray sandy shale	4.00
Hard greenish gray sand rock............................	2.00
Light gray coarse grit	7.00
Gray clay	0.20
Light gray coarse grit...............	3.00
Gray clay	0.10
Gray pebble rock, with dark gray and blackish	
quartzite pebbles	4 00

	Feet.
Greenish gray soft sand rock	0.50
Gray pebble rock, with dark gray and blackish quartzite pebbles	1.50
Greenish gray shale...............................	3.00
Bad *coal* much mixed with earth, (E 444 ba)	1.00
Hidden	7.00
Greenish gray sand rock (E 444 d), perhaps	5.00
Hidden	50.00
Limestone balls	1.00
Greenish gray shales	4.00
Limestone balls...............................	1.00
Greenish gray shale...............................	2.00
Limestone	1.00
Greenish gray shale...............................	4.50
Bad *coal* much mixed with earth (E 445 h)....0.20	
Greenish gray shale.............................2.00	
Bad *coal* much mixed with earth.........,......0.20—	2.40
Greenish gray shale ....,.....,...........	5.00
Limestone balls.............................,....	1.50
Limestone	0.50
Hidden	25.00
Bad *coal*, (E 445 j)..........	0.90
Greenish gray sandy shale, perhaps....................	15.00
Hidden	16.00
Hard greenish gray sand rock.........................	1.00
Soft greenish gray sand rock	0.30
Gray pebble rock with dark gray and blackish quartzite pebbles (B 929 ta).................	2.00
Greenish gray coarse sand rock	4.00
Hidden	10.00
Bad coal (B 926 f)	3.00
Greenish gray sand rock	3.00
Coal slate	1.00
Greenish gray sand rock	2.00

	Feet.
Limestone	1.50
Hidden	5.00
Hard coarse greenish gray sand rock, very hard at bottom, (B 926 r) and (B 889 oa), in all perhaps	20.00
Hidden	3.00
Greenish gray shale (B 929 q), perhaps	15.00
Hidden	26.00
Soft fine gray clay	0.30
Brownish *coal*, (B 889 y)..............0.60	
Coal slate..............0.50	
Coal..............0.40—	1.50
Hidden	100.00
Coal (E 420 ba)	0.15
Coal slate	0.15
Greenish gray shale	1.00
Hidden	100.00
Greenish gray sand rock	0.50
Greenish gray shale	1.50
Bad *coal* much mixed with earth (E 420 d)...0.10	
Limestone balls.............. 0.60	
Bad *coal* much mixed with earth..............0.10—	0.80
Greenish gray shale	1.00
Hidden	40.00
Greenish gray shale	0.50
Bad *coal* much mixed with earth, (E 421 s)	0.10
Greenish gray shale	0.50
Hidden	4.00
Soft brown shale	1.00
Bad *coal* much mixed with earth..............2.50	
Good *coal* (E 425 h)..............1.50—	4.00
Greenish gray sandy shale	2.00
Hidden	14.00
Bad *coal*..............1.80	
Good *coal* (E 421 i)..............1.00—	2.80

	Feet.
Greenish gray sand rock	1.00
Hidden	27.00
Greenish gray shale, perhaps	15.00
Limestone (B 926 p)	1.50
Greenish gray sand rock	3.00
Hidden	3.00
Greenish gray sand rock, perhaps	10.00
Coal slate (B 926 q), about	15.00
Greenish gray sand rock, perhaps	20.00
Hidden	40.00
Green sand rock, exposed perhaps	2.00
Greenish gray shale	1.20
Fine grained sand rock, (B 927 z)	2.00
Greenish sand rock, exposed perhaps	2.00
Hidden	10.00
Coal slate, exposed perhaps	2.00
Bony *coal* 2.50	
Good COAL, (U 607 pc), LOWER BED....1.50—	4.00
Gray shale, exposed perhaps	1.00
Hidden	3.00
Soft clay, exposed perhaps	1.00
Bony *coal* 1.20	
Good *coal* 1.00	
Bony *coal* 0.60	
Good *coal*, (U 607 pb) 1.20	
Bony *coal* ....1.50—	5.50
Coal slate	5.00
Bad *coal*	1.50
Hidden	28.00
Light brown shaly sand rock	1.00
Bony *coal*, (U 607 pa)	1.80
Hidden	45.00
Sand rock with fossil shells (U 607 o), exposed perhaps	12.00

	Feet.
Hidden	32.00
Coal slate (E 423 r)..	4.50
Greenish gray shale	2.00
Hidden	10.00
Coal slate (E 423 sa)	2.00
Hidden	11.00
Greenish gray shale..	1.00
Coal slate	4.00
Greenish gray shale..	6.00
Bad *coal*, (E 423 v)..	0.20
Greenish gray shale..	1.00
Hidden	20.00
Greenish gray shale.....	2.00
Greenish gray sand rock (E 423 wa)................	2.00
Hidden	15 00
Greenish gray sand rock, (E 423 x)..................	7.00
Hidden	35.00
Greenish gray hard sand rock (E 424 z), perhaps....	5.00
Hidden	40.00
Greenish gray hard sand rock, (E 424 b), perhaps...	5.00
Hidden	8.00
Coal slate (E 424 c), perhaps...........................	1.20
Hidden	5.00
Greenish gray shale..	1.50
Coal slate, (E 424 d)	0.80
Greenish gray shale..	5.00
Hidden	30.00
Sand rock with fossil shells (U 607 la), exposed perhaps	12.00
Hidden	45.00
Greenish gray coarse sand rock (E 424 g), perhaps..	10.00
Hidden	100.00
Greenish gray sand rock (E 425 k).................	7.00
Hidden	50.00

	Feet.
Greenish gray shale (E 425 l), perhaps	5.00
Hidden	12.00
Greenish gray sand rock (E 425 m), perhaps	5.00

3461.03

4. *Coal.*—Aside from the small amount of limestone shown in the section, coal is the only mineral of economical importance within the survey. Some of the coal beds although thick seemed to the assistants who made the survey to be of poor quality (at least at some of the exposures); and so have not been counted on as workable at present. Two beds, however, seem satisfactory both in thickness and quality; namely, the bed of (B 884 k) and (Y 663 ba) which averages at those two places 6.30 feet; and the bed of (U 607 pc), (U 622 va), (U 623 b) and (B 927 w), averaging 3.22 feet at the four places. The former (upper) bed, to be quite safe from exaggeration, has been reckoned at only five feet in thickness and the lower bed at only three feet.

The outcrop of the lower workable coal bed runs around the very top of the main saddle or anticlinal already described; and is chiefly high up on the hills on either side of Ôtakisawa in the middle part of its course. The outcrop forms a similar but much smaller oval shape on the corresponding part of the Nagatakisawa hill sides; and reappears in somewhat similar shape but on a larger scale on either side of the Bibai. It probably recurs again in crossing the northeasternmost edge of the survey towards the Sankebibai.

The outcrop of the upper bed is far outside that of the lower bed though rudely parallel to it; and crosses the survey near its southern end in a north-easterly direction. Then on the north of the main saddle it crosses north-easterly the Bibai river in the lower part of its course within the survey ; and has another nearly parallel out-crop half a mile further north on the southern side of the opposite saddle.

The space that is filled by the bed has been measured by means of the map ; and taking account of the steepness of the dip and reckoning a cubic yard to the ton the weight of the coal has been calculated. The upper bed is found to underlie eighty-five acres at a higher level than the lowest natural drainage of the bed within the tract (about 1,275 feet above the sea in the southern part of the tract and 235 feet above the sea on the Bibai), and to contain therein 700,000 tons. Also within 500 feet below that drainage level there are found to be 295 acres or 2,500,000 tons; making 380 acres or 3,200,000 tons above the 500 foot level. Below that level there are besides within the limits of the map probably 230 acres or 2,000,000 tons more, at a depth of not more than 1,300 feet below water level ; making 610 acres or 5,200,000 tons in all.

Of the lower bed in like manner 465 acres within the survey, or 2,360,000 tons are found to be above water level (about 300 feet above the sea); and within 500 feet below water level there are 675 acres or 3,500,000 tons; making in both portions together 1,140 acres or 5,860,000 tons. Besides that there are probably below the 500 foot level 1,600 acres or 8,000,000 tons of the bed within the tract and at most some 3,000 feet below the lowest natural drainage level; making in all 1,740 acres or 13,860,000 tons of the bed.

Taking both beds together, then, there would be within the survey 3,060,000 tons above water level, all lying within 550 acres; and 6,000,000 tons within 500 feet below water level and contained within 970 acres; or in both portions together 1,520 acres or 9,060,000 tons. Adding also the portions below the 500 foot level (or

10,000,000 tons, we find the whole amount of good workable coal within the tract to be 19,060,000 tons contained within 2,740 acres. Some portions of the lower bed underlie a part of the upper bed so as to lessen the number of acres in uniting the two.

There are besides several beds of two feet or more, one even 7.7 feet in thickness, that may be considered as likely to be workable before the large amount already given shall be wholly worked out. The assistants who made the survey, however, considered the thicker of those unreckoned beds to be of inferior quality, and it would perhaps be safest not to take them as quite workable at present ; and only to count on the nineteen millions of tons already mentioned.

5. *Shipments.*—There are at present no roads whatever nor navigable streams within the survey, and the construction of roads would be probably a little difficult, though not impossible, in the narrow, crooked, steep valleys of the Ôtakisawa and Nagatakisawa ; but would be easy enough in the main valley of the Bibai except perhaps in the upper part. The distance by such a road from the coal to the plain of the Ishcari valley would be, say, a mile (14 chô) and thence straight across to the Ishcari river would be about six miles ($2\frac{1}{2}$ ri), and the river would be reached a mile above the mouth of Tomushi. Probably the depth of water up to that point would be about seven feet in ordinary low water. But the great swamp of the Ishcari plain may be too difficult to go straight across with a road. To take a road down the crooked Bibai river would be possible as there is firm ground along its banks ; but the distance to the mouth of the Bibai would be about fifteen miles. The depth of the Ishcari up to that point is called ten feet in ordinary low water. If however a railroad should be built from the Ishcari river (either at Horumuiboto

or at Bibaidap) to the Poronai coal field, the distance to that road from the outlet of the Bibai survey on the eastern edge of the great plain around the foot of the hills would be only about eight miles ($3\frac{1}{5}$ ri), and undoubtedly such a branch road would be very easy to build. It would in fact be, throughout, a part of a line towards the thick beds of fine coal on the Sorachi River ; a line that will almost necessarily be built some day.

A canal might perhaps be dug straight across the Ishcari plain from above the Tomushi to the survey ; but the rise of nearly a hundred feet would require a large number of costly locks. The canal boats or barges could be towed by steamers down the Ishcari River to its mouth ; but the repeated breaking of bulk in loading from the mine railroad into the canal boats and from them again into larger vessels at Ishcari would be a source of some expense and of injury to the coal. It is barely possible that the Bibai river might be converted into a canal ; but in addition to the number of locks that would be required for a still greater fall than that of the canal just suggested, there would be a much greater length of navigation, more even than the fifteen miles above mentioned, because in them the extremely numerous smaller windings of the river are left out of account.

6. *Map.*—The map of the rough survey (dated 19th April, 1876) is on a scale of $\frac{1}{5000}$ and has a number of unexaggerated cross sections on the same scale showing the structure of the rock beds and their dip in several parts of the survey ; likewise a section on a scale of $\frac{1}{1000}$ showing the beds so far as yet known through a thickness of 3461 feet, made out by a combination of all the cross sections and of the strike lines of the map ; likewise two small maps to show the general position of the survey ; and the whole covers a space of 4.75 feet by 2.60 feet. The shape of the surface of the ground is shown

on the main map by contour lines ten feet apart
in level; and the position and shape of the two
workable coal beds are shown in like manner from the
outcrop down to a depth of 500 feet below water level
by means of (broken) contour lines one hundred feet apart
in level and the outcrops are marked besides. The details
of the map are in short the same as those of the Sankebibai and Makumbets surveys and of others already published; but the present map is much larger. As the mode
of drawing the two sets of contour lines and their shape
are so different the fact that both sets are on the same
map gives rise to no confusion; and the place and form of
the coal beds and the direction of the axes are made
very clear.

The surveying was done in 1874 by Messrs. T. Yamauchi, T. Kuwada, T. Kada, I. Ban, E. Yamagiwa and
S. Nishiyama. A main line (not straight throughout)
was run with transit and chain, and side lines, at right
angles or thereabouts, with prismatic compasses and
pacing. The two directions at right angles with each
other were about those of the north-easterly direction of
the axes and the dips across it. Stakes 200 feet apart
(as in the other surveys) were levelled with hand levels.
The surveyor of each line is pointed out on the map.
He also plotted his own line, and the whole was united
into one map by Messrs. Yamauchi and Kada. The geological observations were made chiefly by Mr. Yamauchi.
The geological mapping and sections were done by myself
with the aid of Messrs. Kada and Ban; and Mr. Kada
also lettered and finished the whole map under my own
eye.

I have the honor to be, Sir,
Your most obedient Servant,
BENJ. SMITH LYMAN,
Chief Geologist and Mining Engineer.
Miyôhôji, Echigo;
August 31st, 1876.

GEOLOGICAL SURVEY OF HOKKAIDO.

REPORT

ON THE

NUPPAOMANAI

SURVEY OF 1874;

ACCOMPANIED BY A GEOLOGICAL AND TO
POGRAPHICAL MAP OF THE   SURVEY
AND A TOPOGRAPHICAL MAP OF A
LINE FROM KAWANAI TO THE
ISHCARI RIVER;

BY

BENJAMIN SMITH LYMAN,

CHIEF GEOLOGIST AND MINING ENGINEER.

TOKEI:
PUBLISHED BY THE KAITAKUSHI,
1876.

GEOLOGICAL SURVEY OF HOKKAIDO.

REPORT ON THE NUPPAOMANAI SURVEY OF 1874; ACCOM-
PANIED BY A GEOLOGICAL AND TOPOGRAPHICAL MAP
OF THE SURVEY AND A TOPOGRAPHICAL MAP OF A
LINE FROM KAWANAI TO THE ISHCARI RIVER; BY
BENJAMIN SMITH LYMAN, CHIEF GEOLOGIST AND
MINING ENGINEER.

1.—SITUATION.
2.—LAY OF THE LAND.
3.—GEOLOGY.
4.—COAL.
5.—SHIPMENTS.
6.—MAPS.

HIS EXCELLENCY K. KURODA,

Kaitakuchokuwan.

SIR:

I have the honor to make you the following report on
the Nuppaomanai Coal Survey of year before last.

1. *Situation.*—The survey covers an irregular space
from south-west to north-east through the diagonal of a
square that is nearly three miles and a half (a ri and a
third) on a side, but broad towards the south-west and
tapering off narrowly to the north-east; and containing in.
all 2,410 acres (2,916,000 tsubo) or 3.8 square miles. The
southernmost edge rests upon the Ikushibets River (a
branch of the Horumui) at the mouth of the Nuppaomanai

Brook. The survey on the north-east touches the southern edge of the Bibai survey; on the south touches the Rail Road Survey of last year; and on the north-west adjoins the survey of the line run year before last to the Ishcari River.

2. *Lay of the Land.*—The Nuppaomanai Brook traverses the whole eastern edge of the survey from the very north-eastern corner, where a high ridge divides it from the Bibai valley. Several small branches join the brook from the north and have their valleys almost wholly within the survey. The remaining western part of the survey is filled by the head waters of the Kawanai brook which flows south-westward, and is said by the Ainos to end in the great swamp of the Ishcari valley without reaching either to the Ikushibets or to the Chipturaship, a branch of the Bibai.

For about a thousand yards northward from the southern edge of the survey the land is very flat, and about 150 feet above the sea; then it rises fifty feet to a terrace, the brow of which is however towards the west cut down into a gentle slope; then at nearly a mile from the southern edge of the survey steep hills begin which fill up the whole remaining space, except narrow bottoms, scarcely 200 feet wide, along the lower part of the Nuppaomanai and Kawanai Brooks. Towards Kawanai the hills are only 550 to 650 feet high above the sea but north-eastward are higher and higher until at the upper corner of the survey they are 1,800 feet high. The land is everywhere covered with forest and wholly uninhabited by men.

3. *Geology.*—The rock beds are very much contorted, and at first it was difficult to make out any probable explanation of the extremely various dips. For half a mile or more from the north-east corner of the survey, however, the comparatively regular and simple structure of the southern part of the Bibai survey with a north-east and

south-west saddle axis and comparatively gentle dip seems to continue, bringing higher and higher rocks to the surface, so that in fact the workable coals of the Bibai and Sankebibai surveys do not reappear at all in the Nuppaomanai survey. Then begins a region of steep dips in many different directions, that fill up most of the remainder of the survey except perhaps the southern flat ground, where the rocks are wholly concealed, and where perhaps rocks of the Toshibets Group (later than the other rocks of the survey, the coal bearing group) lie below, since they seem to be exposed at Poronaibuto close by. The confused appearance of the dips just spoken of is caused by a crushing together of the beds that has produced on the north-east a narrow steep basin, then a narrow sharp saddle, then another narrow basin and another rather broader saddle with their axes all running north and south, crossing the principal axial direction of north-east and south-west which is seen in the Bibai survey on one side and in the Poronai survey on the south. Even in the north-western edge of the Nuppaomanai survey near the beginning of the Ishcari Line is found a steep sided rock saddle with its axis running north-east and south-west and between that and the four north and south axes just mentioned is still another very small but very steep saddle with two little branches north-eastward which appears to have its axis running mainly north-easterly and south-westerly. Most of the observed dips therefore are extremely steep, except in the north-eastern corner of the survey, but even there are often more than thirty degrees and seldom less than twenty degrees.

The rocks of the survey, then, belong, so far as known, to the Horumui (or Brown coal) Group, and are almost wholly above those of the Bibai survey, and in the upper part correspond to those of the Poronai survey. The following section (from above downwards) has been made out by a combination of all the field observations through means of the geological and topographical map ; and the closeness of its agreement above and below with the Poronai and Bibai sections, which were worked out for the most part quite independently, very strongly confirms the explanation of the structure just given :—

	Feet.
Greenish white micaceous sand rock	3.00
Weathered brown sand rock	6.00
Flinty shale containing clay and lime, (K 942 l)	7.00
Hidden	7.00
Bluish gray sand rock, weathering brown (U 358 i), exposed perhaps	3.00
Hidden	36.00
Soft clay, exposed perhaps	3.00
Soft *coal*, (U 360 q)	0.40
Soft clay	0.20
Soft *coal*	0.20— 0.80
Soft clay, exposed perhaps	3.00
Hidden	125.00
Bad *coal*, (E 323 t), probably the Poronai No. VII coal imperfectly exposed	1.00
Brownish gray sandy clay, exposed perhaps	2.00
Hidden	15.00
Grayish white shale, (K 947 q), exposed perhaps	2.00
Hidden	8.00
Yellowish white sand rock, (K 948 z), exposed perhaps	2.00
Hidden	25.00
Grayish white shale, (K 947 s), exposed perhaps	2.00
Hidden	20.00
Yellowish white fine sand rock, (K 948 v), exposed perhaps	2.00
Hidden	38.00
Gray clay shale, (K 948 u), exposed perhaps	3.00

	Feet.
Hidden	50.00
Thin traces of *coal*, (U 330 k), probably the outcrop of the Poronai No. VI coal	
Bluish gray sand rock, exposed perhaps	2.00
Hidden	75.00
Brownish gray shale, exposed perhaps	2.00
Bad *coal*, (E 322 m)	1.00
Greyish brown clay	1.00
Reddish brown clay, exposed perhaps	3.00
Hidden	35.00
Greenish brown clay	0.40
Hard and good *coal* (U 330 i)	0.56
Greenish brown clay	0.40
Greenish hard clay contain *coaly* fossils	4.40
Bluish gray hard clay	5.00
Blue grit, exposed perhaps	2.00
Hidden	3.00
Yellowish gray sandy shale, (K 950 f), exposed perhaps	3.00
Hidden	27.00
Grayish sand rock, (K 950 g), exposed perhaps	3.00
Hidden	110.00
Hard brown clay, exposed perhaps	2.00
Bad *coal*, (E 327 ea)	0.30
Hard brown clay, exposed perhaps	2.00
Hidden	62.00
Rotten coal slate, (E 327 ca), perhaps	0.50
Hidden	5.00
Hard gray clay, (E 327 a), perhaps	4.50
COAL (locally much crushed and poor)	1.00
Clay	0.30
COAL (locally much crushed and poor), (Y 552 c) UPPER BED	4.20
Light gray clay	1.50

	Feet.
Coal ....2.00—	9.00
Gray shale, (U 377 u) exposed perhaps...............	3.00
Hidden	30.00
Blackish clay...	1.00
COAL pretty good (but locally rotten), (U 329 d), MIDDLE BED	3.50
Blackish clay...	0.30
Poor dirty *coal* or bituminous shale, rotten.	3.70
Dark gray clay..	0.50
Black *coaly* matter much mixed with layers of clay, (U 329 e)	31.00
Whitish clay, (Y 576 b)..	10.00
Hidden	20.00
Hard gray clay, exposed perhaps.....................	1.50
Coal in pieces, mixed with coal slate, (Q 441 b)...	.300
Hard gray clay, exposed perhaps.....................	3.00
Hidden	13.00
Greenish sand rock	15.00
Coal mixed with layers of clay, (B 801 f)...4.00	
Bad *coal*.. ....1.00—	5.00
Bituminous clay	1.00
Hidden	52.00
Greenish gray shale, exposed perhaps...............	2.50
Coaly matter, (B 801 c)	1.20
Soft mud and fragments of shale.......................	0.20
Coaly matter	1.00
Greenish gray shales, exposed perhaps.	1.50
Hidden 145.00	
Sand rock, exposed perhaps...........................	1.00
Soft clay	0.30
Soft COAL 2.10	
Hard COAL, (K 952 p), LOWER BED.......... 2.60—	4.70
Hidden	35.00
Sandstone, (Y 634 m), exposed perhaps...............	1.50

	Feet.
Hidden	25.00
Soft sand rock, exposed perhaps............	1.00
Bad coal, (Q 440 d)..................................	0.15
Hard dark gray clay, exposed perhaps	2.50
Hidden	30.00
Sandstone, (Y 634 n), exposed perhaps...............	2.50
Hidden	240.00
Sand rock................................... ?	
Gray clay..... .... ?	
Poor coal, (Q 438 g)...........................1.50	
Coal slate .... ? —	4.75
Hidden	115.00
Shales, exposed perhaps	3.00
Bad *coal*, (U 513 w)	0.50
Shales, exposed perhaps	3.00
Hidden	105.00
Soft clay, exposed perhaps	1.50
Hard *coal* mixed with a little clay, (U 507 r) 2.30	
Good *coal*, but a little soft ...'................0.50—	2.80
Gray shales, exposed perhaps..........................	2.50
Hidden	7.00
Soft clay, exposed perhaps	2.50
Hard *coal* mixed with a little clay.............0.35	
Soft clay......... ....0.20	
Good *coal*, (U 507 l)..........................0.70—	1.25
Gray shales, exposed perhaps...........................	2.50
Hidden 180.00	
Coal, (E 406 ab)................................	0.35
Light bluish sand rock, exposed perhaps	2.50
Hidden.....................................	35.00
Good *coal*...0.25	
Sandy rock................................0.15	
Bad *coal* (B 869 bb)............................0.35—	0.75
Coal slate ....,............................	0.35

	Feet.
Hidden	28.00
Hard brown sand rock	2.00
Rather hard brown sandy clay	2.00
Reddish brown bad *coal* 0.30	
Soft gray clay 0.10	
Coal 0.10	
Soft gray clay 0.30	
Coal 0.10	
Reddish brown soft clay 0.20	
Hard gray clay 1.10	
Soft whitish brown clay 0.30	
Hard dark gray clay 0.50	
Coal 0.10	
Soft dark gray clay 0.30	
Coal 0.60	
Soft dark gray clay 0.20	
Bad *coal* 0.45	
Soft light gray clay 0.10	
Coal 0.15	
Soft light gray clay 0.05	
Bad *coal*, (Q 532 a) ....0.80—	5.75
Hard reddish brown clay, exposed perhaps	3.00
Hidden	15.00
Reddish gray coarse grit, (Y 629 jb), exposed perhaps	3.00
Hidden	30.00
Reddish gray sandy clay, exposed perhaps	1.00
Soft gray clay	0.10
Coal 0.20	
Soft whitish gray clay 0.10	
Coal 0.70	
Reddish brown fine sand rock 0.30	
Coal 0.10	
Reddish bad *coal*, (Y 646 fa) ....0.45—	1.85

	Feet.
Whitish hard gray clay, exposed perhaps	1.00
Hidden	25.00
Soft gray clay............ ...., ...	0.10
Coal, (Q 534 e)..	0.30
Hard gray clay, exposed perhaps................	3.00
Hidden	30.00
Hard gray clay, (U 383 a), exposed perhaps	3.00

2040.06

4. *Coal.*—The only mineral of any economical value within the survey is the coal; and of that three workable beds are exposed. The Lower Bed was found at only one point (K 252 p), and is the same as the Poronai No. I, or Bottom (workable) coal. Although it is 4. 7 feet thick at the Nuppaomanai exposure it averages at nine places in Poronai only 3.67 feet in thickness; or taking all ten places together 3.77 feet. Supposing it to have become generally thicker towards Nuppaomanai it may be taken as four feet and a half thick.

The Middle Bed is that of (U 329 d) and is the same as the Poronai No. IV coal (that of (L 602 ba)). It is exposed at ten places in Nuppaomanai with a thickness of from 2.5 feet to 5.8 feet of coal, and an average of 3.96 feet; but at five more places in Poronai, of 5.79 feet; or an average in all together of 4.75. As some of the Nuppaomanai exposures are perhaps imperfect, it is probably safe to count even there on an average thickness of four feet and a half.

The Upper Bed is only exposed in Nuppaomanai at (Y 552 c) with a thickness of 7.2 feet of coal; but it corresponds to the Poronai No. V coal (or that of (L 602 cb)), which is exposed at three places in Poronai with an average there of 5.1 feet of coal; giving an average of 5.87 feet for all four places. In spite, therefore, of the greater thickness at Nuppaomanai it would perhaps be safest not to count upon more than six feet there.

The Poronai No. II coal (averaging 3.56 feet at four places) has not been seen at all at Nuppaomanai, but may be discovered at some future time. It lies about 80 feet above the Poronai No. I. The Poronai No. III Bed (averaging there 3.33 feet of good coal) seems at Nuppaomanai to be the coal of (B 801 c) with 2.2 feet of "coaly matter"; perhaps an imperfect exposure. The Poronai No. VI (or (L 576 i)) Bed (averaging there 4.3 feet of good coal) seems to correspond to the "thin traces of coal" at (U 330 k) ; and perhaps a more thorough opening there might bring a workable bed to light. The Poronai No. VII Bed (averaging 4.7 feet of coal at two places there) seems at Nuppaomanai to be perhaps the bed of (E 323 t), "one foot of bad coal" and that of (K 943 p) "four feet of bad coal"; and if the former is a complete exposure of the bed and the quality has not been underestimated (by the assistants) at either place, the bed is quite unworkable for the present in Nuppaomanai. It is however not at all improbable that some of these four beds found so satisfactory at Poronai may prove on further exploration to be workable at Nuppaomanai also.

Owing to the very steep dips at Nuppaomanai and the fact that most of the coal exposures are either right at rock saddles or close by them, the coal (except the Lower Bed) has a crushed and, at the outcrop, very rotten or much weathered and dirty appearance. As the beds, however, are the same as those at Poronai where the dips are gentler, and owing to the shortness of the distance (scarcely a league) there can really have been originally no great difference in the beds themselves; it is perhaps best to judge by their appearance at Poronai of their character as it will be found in the less disturbed portions of Nuppaomanai and at a good depth below the weathering influences of the outcrop. Assays have been made and published of samples from the two upper beds as found at Poronai Nos. IV and V.

The outcrops of the three Nuppaomanai beds owing to to the steepness of the dips are for the most part but little affected in respect to direction by the varying height of the ground, and for the same reason are nearly parallel to each other and generally not far apart. They curve round the north-west anti-clinal in an oval shape ; and in somewhat similar form encircle the summits of the small middle saddle and its two little north-eastern branches ; and in irregular ess shape pass around the two basins and eastern saddle of the group of four north and south axes ; ending with a nearly straight course northerly to the edge of the survey, at a distance of nearly a mile from the extreme eastern edge. The other Poronai beds would also have their outcrops in general parallel and near to these and would therefore probably be not difficult to find now in searching along the bottom of the narrow valleys with the help of the map, which shows the place of the three beds.

The same map enables us to measure the extent of those beds within the survey, and taking into account their thickness and the steepness of the dip to calculate their cubic contents, and reckoning a cubic yard to the ton to give the number of tons of each above a given level.

Of the Upper Bed, then, there are 148 acres (179,000 tsubo) or (at a thickness of six feet) 2,205,000 tons above the lowest natural drainage level, which is for the north-western and middle saddle, about 220 feet above the sea, and for the north and south basins and saddles about 370 feet above the sea. Within 500 feet below those levels there are 235 acres (284,000 tsubo) or 3,765,000 tons; making in all above that depth 383 acres (463,000 tsubo) or 5,970,000 tons. Besides that at a greater depth but less than 4,000 feet below sea level there are probably 650 acres (785,000 tsubo) or 11,200,000 tons; making the whole workable amount of the bed to be 1,033 acres (1,248,000 tsubo) or 17,170,000 tons.

The Middle Bed in like manner underlies 150 acres (182,000 tsubo) above the lowest natural drainage level (about 220 feet above the sea at the two western saddles and 370 feet above the sea at the eastern ones) and at a thickness of four feet and a half would amount there to 1,430,000 tons. Within 500 feet below drainage there are 243 acres (293,000 tsubo) or 2,440,000 tons; making in all above that depth 393 acres (475,000 tsubo) or 3,870,000 tons. In addition, at a lower depth but not more than 4,000 feet below sea level there are probably 800 acres (970,000 tsubo) or 8,200,000 tons; making the whole amount of workable coal in the bed within the limits of the survey to be 1,193 acres (1,445,000 tsubo) or 12,070,000 tons.

Of the Lower Bed there are likewise above the lowest natural drainage level (about 240 feet above the sea at the two western saddles, and about 600 feet above the sea at the eastern ones) 67 acres (81,000 tsubo) or 560,000 tons, reckoning the thickness at four feet and a half. Within 500 feet below drainage there are 153 acres (184,000 tsubo) or 1,440,000 tons; making in all above that depth 220 acres (265,000 tsubo) or 2,000,000 tons. Moreover, between that depth and 4,000 feet below sea level there are probably 930 acres (1,125,000 tsubo) or 9,100,000 tons; making the whole amount of the workable coal of this bed within the survey to be 1,150 acres (1,390,000 tsubo) or 11,100,000 tons.

All three beds together, then, have 4,195,000 tons above natural drainage level; 7,645,000 tons within 500 feet below that level; or 11,840,000 tons in both portions together. Besides that, between the lower (500 foot) level and 4,000 feet below the sea there are probably 28,500,000 tons ; making the whole amount of workable coal of the survey, in the three beds, to be 40,340,000 tons. The numbers of acres of the different beds are not to be simply added together like the tons, because some portions of one bed in many places overlap portions of another bed.

5. *Shipment.*—There are no navigable streams nor roads at present within the limits of the survey. The Ikushibets which just crosses the southern edge of the tract is scarcely navigable for log canoes during a great part of the year ; and probably could not be made navigable for coal barges or canal boats without far too great expense for locks in a fall to the Ishcari at Horúmnibuto of about a hundred feet. Owing also to the remarkable crookedness of the stream the distance would be very great. The nearly equal height of fall and the number of locks required would likewise make it probably too costly to dig a canal more directly across the great swamp to the Ishcari River near the mouth of the Bibai ; although the distance from one of the upper bends of the Ikushibets would be only seven miles and a half (3 ri) and from another bend lower down only five miles and a half ($2\frac{1}{4}$ ri); and the distance would be even two miles ($\frac{3}{4}$ ri) shorter, to a certain branch of the Bibai, that is nearly in the straight line. In case, however, a canal should be dug from the Ikushibets an additional two miles either partly or wholly of rail road would be needed in order to bring the coal to the Ikushibets.

It is possible that a rail road might be built across the swamp to the Ishcari near the mouth of the Bibai, and the whole length from the coal would be about eleven miles and a half. Such a road would be part of some future line directly across the swamps west of the Ishcari to the town of Ishcari on the sea shore; a line that would make the shipment of the coal quite independent of the ice that closes the river for four months every winter. But the softness of the swamp may possibly prove too great an obstacle to the building of such a road and make it preferable to build one down the valley of the Ikushibets and Horumui to Horumuibuto. In that case it would be shortest and probably best to cross the Ikushibets near Poronaibuto, and join the line for a rail road surveyed last summer. The distance from the coal to Horumuibuto would then be about $17\frac{1}{2}$ miles (seven ri) about the same as from the Poronai coal to the same point. The road above Poronaibuto would for the first mile be even easier to build than the Poronai road, since the ground is much flatter ; but above that the valleys became very narrow, and the building of a steam railroad would probably be difficult though not impossible. The coal of the north-western saddle, as it is in the Kawanai valley, would have to be reached by a branch road through a low gap there is in the hills. As the dips are so steep in the Nuppaomanai field and the coal so much crushed, it will be less desirable to work (other things being equal) than the Poronai coal. The Nuppaomanai coal will also suffer more in transportion from its greater tendency to separate into small bits.

6. *Maps.*—The surveying was done in 1874 by Mr. T. Yamauchi, Assistant Geologist, aided by Messrs. T. Kuwada, T. Kada, I. Bau, E. Yamagiwa and S. Nishiyama, Assistant Geologists. A main line was run with transit and chain first north-westerly (part of the line to the Ishcari River) across the south-west end of the survey and then at right angles with that through the middle of the survey to its north-eastern corner ; and side lines were run with prismatic compass and pacing. Stations commonly two hundred feet apart, were levelled with hand levels. Each surveyor plotted the lines and drew the

contours of his portion of the field and the whole was united
into one map by Mr. Yamauchi with the aid of the others.
Last summer I was able myself to visit the coal exposures
near the small middle rock saddle, and take the dips and
measurements of thickness there, which had been made
difficult by the contortions of the beds. The other geological observations were made chiefly by Mr. Yamauchi.
The geological mapping and sections were done by myself
with the aid of Mr. Nishiyama; who also, under my eye,
traced, lettered and finished the final copy with some help
towards the end from Mr. Kada. The map (dated 6
May, 1876) is on a scale of $\frac{1}{5000}$ and has numerous unexaggerated cross sections to show the structure and dip
at different points; also a columnar section on a scale of
$\frac{1}{1000}$ to show the different beds described above besides
numerous other exposures of the same beds in parallel
columns; and there is likewise a small map to show the
general position of the field. The whole map, therefore,
covers a space of 3.53 feet square. The main map like
those of our other surveys shows the shape of the ground
by contour lines ten feet apart in level. The shape of
the coal beds is shown by contour lines of a similar
nature but far more regular in shape, and thereby
readily distinguishable, as well as by the fact that they
are drawn as broken lines ; and they are 100 feet apart
in level and reach only to 500 feet below water level.
The outcrops of the three beds are also marked by short
crosshatched lines. The surveyor of each line is indicated
and the staked stations distinguished by different marks,
so that any point may readily be found on the ground.
The details of the map are in general the same as those of
our other maps of the same scale.

Besides the map of the main survey there is a topographical map on the same scale, dated 7th June, 1876,
and 9.44 feet long by 1.68 feet wide, that shows a

line run from Kawanai at the south-west corner of the main survey north-westerly to the Ishcari River. The land is almost wholly a swampy alluvial plain, and no rock exposures were observed. The coal bearing rocks however probably exist not only in the hill sides at the south-eastern end of the survey but also under the alluvium throughout the rest of the map.

I have the honor to be,

Sir,

Your most obedient servant,

BENJ. SMITH LYMAN,
Chief Geologist and Mining Engineer.

Miyôhôji, Echigo ;
2nd September, 1876.

GEOLOGICAL SURVEY OF HOKKAIDO.

REPORT

ON THE

PORONAI

COAL SURVEY;

ACCOMPANIED BY A GEOLOGICAL AND
TOPOGRAPHICAL MAP OF THE COAL
LANDS AND BY A TOPOGRAPHICAL
MAP AND A PROFILE SECTION OF
A SURVEY FOR A RAIL ROAD
LINE TO HORUMUIBUTO;

BY

BENJAMIN SMITH LYMAN,

CHIEF GEOLOGIST AND MINING ENGINEER.

TOKEI:
PUBLISHED BY THE KAITAKUSHI,
1876.

GEOLOGICAL SURVEY OF HOKKAIDO.

REPORT ON THE PORONAI COAL SURVEY; ACCOMPANIED BY A GEOLOGICAL AND TOPOGRAPHICAL MAP OF THE COAL LANDS AND BY A TOPOGRAPHICAL MAP AND A PROFILE SECTION OF A SURVEY FOR A RAIL ROAD LINE TO HORUMUIBUTO; BY BENJAMIN SMITH LYMAN, CHIEF GEOLOGIST AND MINING ENGINEER.

1.—SITUATION.
2.—LAY OF THE LAND.
3.—GEOLOGY.
4.—COAL.
5.—SHIPMENT.
6.—MAP.

HIS EXCELLENCY K. KURODA,
                    Kaitakuchokuwan:

SIR:

I have the honor to make you the following report on the Poronai Coal Survey, made chiefly in 1873 and 1874.

1. *Situation.*—The survey has in general a nearly square shape with some large notches in its sides and is, about two miles (four-fifths of a ri) long from north to south and from east to west, and amounts in all to 1,740 acres (2,105,000 tsubo) or 2.7 square miles. It covers nearly the whole of the head waters of the Poronai valley and reaches to within about a mile of its mouth on the Ikushibets a large branch of the Horumui (perhaps strictly the main stream).

2. *Lay of the Land.*—The Poronai Brook flows out at the north-western corner of the survey ; and the valley of its main branch, with a westerly course, fills the northern part of the tract ; but from a large south-western branch near the middle of the great square several smaller branch valleys radiating to the south-west, south and south-east fill up the remainder. There is commonly a strip of comparatively flat land about 300 yards wide along the brook up to the main forks ; and from there to Hayakawa village (so called) about one hundred yards wide; but the rest of the streams generally flow with a rapid fall at the bottom of steep hill sides on either hand. The main brook on leaving the tract is about 150 feet above the sea ; but the hills, almost everywhere steep, rise in the northern part of the survey to 800 feet, in the central and south-western to 1,000, and on the south-eastern to 1,300 feet. The land is everywhere covered with forest and almost everywhere with bamboo grass (*Arundinaria*) ; and is wholly uninhabited by men. There is however one large bark hut at the principal one of the forks near the centre of the tract and the place has been called Hayakawa village in honor of the cooly who first brought the Poronai coal to the notice of the government, more than three years ago

3. *Geology.*—The geological structure through the greater part of the survey has been very fully made out, and is in general very simple. The rock beds lie in the form of a great saddle with the axis running nearly north-east and south-west ; and plunging towards the north-east, from near the south-west corner of the survey, the highest point of the saddle ; then near the north-eastern corner of the survey rising again in the same direction. The dips are comparatively gentle (45° and less) on the northwestern side of the saddle; but very steep (60° or more) on the other side. The plane of the axis, then, has a very marked inclination towards the north-west, and a line

drawn along it at one level would, owing to varying
steepness of dip, not be exactly parallel to a line upon
it at another level. Such a line moreover would not be
straight ; because the shape of the north-western side of
the saddle is affected by a shallow basin that is impressed
upon it with an axis running about north and south ; and
the axis is impressed to some degree with a like curve, or
a gentle convexity towards the south-east. The same
tendency to a north and south folding of the rock seems
to have produced certain small faults of a few feet or a
few yards, six or seven of which have been found upon or
near the main (north-westerly) branch above Hayakawa
village ; but their extent and dip are not yet known.
The two sets of axes, north and south and north-east and
south-west, are also seen to exist together to a still more
marked degree in the Nuppaomanai survey and can also
be perceived in the Bibai survey where the north-east
and south-west direction is the principal one again. In
the Sankebibai survey still further north, the north and
south direction is the only one yet observed. The structure
in the south-eastern corner of the tract has not yet been
studied out, owing to lack of observations.

The rocks of the survey belong wholly to the Horumui
(or Brown Coal) Group ; and the following is a section of
them (from above downwards) that has been made out
from a combination of all the observations in the field by
means of the geological and topographical map, a much
more correct and complete section than the one I was
able over two years ago to prefix to the chemical report
on our Yesso coals :—

		Feet.
Good *coal*, (B 685 a)	....	1.79
Hidden	....	15.00
Coal traces, (B 682 1)	....	—
Hidden	....	30.00

	Feet.
Coal traces, (B 682 k)..........................	—
Hidden	35.00
Good *coal*...0.35	
Soft poor *coal*..0.15	
Good *coal* (B 670 r)........,.......................1.50	
Soft poor *coal*...0.20	
Coal slate...2.00	
Coal ....0.15—	4.35
Hidden	55.00
Coal traces, (B 682 i)...	—
Hidden...480.00	
Greenish gray sand rock, (B 669 l) with fossils (clams), perhaps......	2.50
Hidden...... ....145.00	
Hard grayish brown clay, (Y 417 ma), exposed perhaps..,..............	3.00
Hidden....................., ...270.00	
Sand rock, exposed perhaps...........	3.00
Coal..•........	2.80
Hidden...170.00	
Bluish gray sand rock weathering brown.............	1.80
Coal (K 745 in)..	1.70
Coal slate..	0.50
Grayish brown soft clay..................................... ..	1.30
Light gray soft clay with a little *coal*................	0.80
Hidden 130.00	
Light gray coarse sand rock.............................	2.00
Good COAL, (Y 411 gb), No. VII............. 1.40	
Bony *coal*...0.90	
Poor COAL...1.40	
COAL mixed with gray clay....................1.80—	5.50
(At (K 746 b) the same bed has 6.60 of "good COAL")	
Coaly clay	0.80

	Feet.
Hidden	75.00
Greenish gray shaly sand rock weathering brown...	8.00
Bean size pebble rock with white pebbles and with	
a 0.5 seam of black coal slate in the middle	0.30
Coal slate	0.20
Same pebble rock..	1.00
Hidden	0.80
Coal (L 577 fa), exposed...............................	0.50
Hidden	29.00
Gray sand rock with fine black specks, weathering	
brown, exposed about	8.00

Bony *coal* 0.05	
Good COAL (L 576 i), No. VI..................4.20	
Bony *coal* 0.05	
Black clay 0.10	
Bony *coal*............. ....0.50	
Rotten *coal* ....0,10—	5.00

Good bluish gray soft fire clay..........	1.50
Hidden....................	86.50
Irregular bed of *coal* (L 576 m).......................	0.80
Hidden	42.00
Greenish gray sand rock weathering brown, ex-	
posed perhaps...............................	8.00
Gray shales, rather soft.................................	1.00

Soft *coal* 0.60	
Bony and soft *coal* mixed.........................0.25	
Good c al with 0.02 of slate at 0.35 from the	
top (L 581 p) 1.00	
Black coal slate with a very little *coal* 0.50	
Blackish soft shales dark gray at top.........1.00	
Bony coal ....0.60—	3.95

Grey soft shales	3.00
Greenish gray sand rock weathering brown, ex-	
posed perhaps	8.00

	Feet.
Hidden	1.50
Rotten *coal* (606 p b), about	1.00
Hidden	73.00
Hard dark gray shales	0.20
Coal (L 606 on)	0.50
Hard dark gray shales	0.20
Hidden	13.00
Blackish shales with 0.10 of poor *coal* in the middle and 0.20 of same at bottom (L 604 kc)	2 00
Dark gray shales, exposed about..	0.30
Hidden	30.00
Coal quite hard (L 581 qa) 1.00	
Black slate 0.02	
Very hard *coal* ... .0.40—	1.40
Dark gray clay	0.20
Dark gray shale bluish at top, exposed .	0.30
Hidden	13.50
Dark gray and blackish clays:	1.20
Crushed *coal* 0.90	
Dark gray and blackish shales................0 50	
Coal poor or crushed (L 604 ka)...............0.95—	2.35
Black coal slate......	0.85
Hard gray sandy shales, exposed	1.00
Hidden	50.00
Blackish coal slate	0.50
Coal 0.75	
Dark gray shale................0.65	
Coal................0.80—	2.20
Blackish coal slate	0.05
Greenish gray sand rock dark at top	0.70
Hidden, about	4.00
Greenish gray sand rock weathering brown, about	1.00
Dark gray and black coal slate	3.40
Coal, poor, soft and tender 0.45	

Feet.

Dark coal slate 0.40
Good firm COAL (L 602 cb), No. V....... ....1.70
Dark gray and black coal slate...... . ...0.45
Poor *coal*, tender.............. ....0.40
Good firm COAL........................ ... .1.50
Dark shale................... ....0.10
Good firm COAL 1.60
Hard blackish coal slate and *coal* mixed......0.30
Very hard COAL....................0.65—   7.55
Blackish coal slate................................   0.20
Soft blackish clay................................   0.30
Light gray clay shale, exposed 0.50
Hidden 19.00
Brown weathered shale, exposed perhaps............   2.00
Black coal slate..................................   0.15
Coal bony 0.65
Blackish coal slate.........................1.20
Coal.......................................0.10
Blackish coal slate.........................0.30
Coal......................................0.25
Blackish coal slate.........................0.05
Coal ....0.05—   2.60
Hard greenish gray sand rock weathering brown...   0.70
Blackish coal slate, perhaps thin streaks of coal......   1.85
Hard dark gray sand rock 1.30
Dark, mostly black, coaly shale 4.70
Coal 0.30
Greenish gray sand rock, weathering brown.........   1.20
COAL 1.60
Black coal slate 3.20
COAL (L 602 ba), No. IV...................5.00—   9.80
Black coal slate, hard 1.50
Hidden 100.00
Coal (L 586 r)..................................   1.66

	Feet.
Dark gray hard shale	0.20
Hidden	50.00
Greenish sand rock weathering brown	2.50
Black coal slate, hard	0.20
COAL very firm (L 586 ta), No. III.	4.00
Dark hard shale	0.05
Hidden	16.00
Greenish gray sand rock	0.80
Black or bluish hard coal slate	2.30
Coal....0.45	
Blackish hard coal slate....1.50	
Coal, (L 587 v)....0.15—	2.10
Black coal slate, hard	0.65
Blackish gray shales, hard	1.25
Black coal slate, hard with trace of *coal*	0.30
Dark gray shales	0.50
Hidden	53.00
COAL very firm (L 587 ba), No. II....3.00	
Black coal slate hard, perhaps partly *coal*...0.35	
COAL....0.20—	3.55
Clay soft, exposed perhaps	0.10
Hidden	20.00
Coal, possibly in very small part bony, (L 587 bb).	2.80
Gray shales, exposed about	0.20
Hidden	25.00
Coaly brown rock	0.20
Coal slate	0.30
Coal (Q 128 l)	0.70
Blackish gray clay	0.40
Hidden	10.00
Coal (L 587 da)	0.55
Black slate	0.10
Hidden	3.00
Dark gray shale, about	0.50

	Feet.
Black slate (L 587 d)..	1.60
Dark gray shales, about..............................	0.70
Black slate, perhaps...	0.30
Hidden	9.00
Coal (T 145 a), No. I.. ....2.00	
Dark blackish coal slate, hard 1.20	
Coal 0.15	
Black coal slate, hard 0.30	
Coal 0.15	
Black coal slate..................................0.05	
Coal very hard........................1.65—	5 50
Dark gray hard shales, blackish on the surface, about..	0.50
Hidden	12.00
Greenish gray sand rock, exposed perhaps	1.00
Hidden...	3.00
Greenish gray sand rock, exposed perhaps...........	1.00

2,239.90

4. *Coal.*—With the exception of some large balls of limestone which seem to contain a certain amount of carbonate-of-iron ore, the only mineral of economical importance is the coal, of which seven workable beds are exposed.

The Bottom (workable) Coal, or No. I, is that of (T 145 *a*), and it is exposed at nine places in Poronai with an average thickness of 3.67 feet of good firm coal. One more exposure in Nuppaomanai raises the average for all ten places to 3.77 feet.

The No. II Coal is that of (L 587 *ba*), which is exposed at four places with an average thickness of 3.56 feet of good firm coal. No exposure of the bed has yet been found outside of Poronai.

The No. III Coal is the one exposed at (L 586 *ta*) and at two other places in Poronai with an average thickness of 3.33 feet of good firm coal. It is also exposed at one place in Nuppaomanai and at one place in the Bibai survey and the average of all five is 3.14 feet of coal.

The No. IV Coal is the one of (L 602 *ba*) of which five exposures have been found at Poronai with an average thickness of 5.79 feet of good coal; and ten exposures at Nuppaomanai with an average for all fifteen of 3.96 feet of good coal; but some of the Nuppaomanai exposures are perhaps imperfect ones.

The No. V Coal is exposed at (L 602 *cb*) and at two other places in Poronai with an average for all three of 5.10 feet of good coal; and at one place in Nuppaomanai, making the average for the four places 5.87 feet.

The No. VI Coal is that of (L 576 *i*) and is exposed also at another place in Poronai with an average for the two of 4.30 feet of good firm coal. The two places are those which were marked last year as good places to begin mining coal, and are near the end of the main railroad line and of its branch. The same bed shows "thin traces" of coal only at Nuppaomai; but has perhaps not been properly opened there yet.

The No. VII Coal is exposed at (Y 411 *gb*) and at another place in Poronai with an average of 4.70 feet of coal that is probably good; and there are two other exposures (perhaps imperfect ones) at Nuppaomanai which reduce the average of all four to 3.80 feet and make the quality still more doubtful.

Specimens of the coal of the Nos. IV, V and VI beds were assayed by Mr. Munroe, and the results were published in his report over two years ago. The true relative position of the different beds had at that time been far less satisfactorily ascertained than it has been now; so that a few words are necessary here to explain

to which of the above mentioned beds the assayed specimens belonged; and some of the principal results of the
assays and analyses may be recalled by the two accompanying tables.

In descending order:

The (L 576 *i*) Coal is the No. VI Bed, and is the same
as that of (L 1,996 *b*).

The (L 602 *a*) Coal is the No. V Bed, and is the same
as that of (L 602 *cb*) and (L 306 *m*).

The (L 602 *ba*) Coal is the No. IV Bed, and is the same
as that of (L 602 *ca*), (L 603 *e*), (L 602 *bb*) and (606
ia), and (L 603 *g*).

ASSAYS.

Name of Coal	Specific Gravity.	Moisture at 140° C.	Volatile Combustible.	Fixed Carbon.	Ash.	Percentage of coke.	Appearance of coke.	Color of ash.
Horunui (L 576 i) ....4.2 ft.	1.284	5.62	39.88	52.32	2.18	54.50	very poor.	Dark umber.
,, (L 602 a) ....5.0 ft.	1.281	5.19	37.51	54.84	2.46	57.30	poor.	Bright orange.
,, (L 602 ba) upper bench..1.9 ft.	1.277	4.20	40.42	51.99	3.39	55.38	good.	Orange yellow.
,, ,, lower bench..5.0 ft.	1.286	4.25	41.26	52.08	2.41	54.49	fair.	,, ,,
,, (L 602 ba) calculated average	1.283	4.24	41.01	52.05	2.70	54.75	fair?	,, ,,
,, (L 602 ca) upper bench..2.0 ft.	1.305	5.03	38.29	53.31	3.37	56.68	fair.	,, ,,
,, ,, lower bench..2.7 ft.	1.304	4.58	40.67	50.45	4.30	54.75	good.	,, ,,
,, (L 602 ca) calculated average	1.305	4.85	39.24	52.17	3.74	55.90	fair.	"Yellow."
,, (L 603 e) ....5.0 ft.	1.282	4.47	40.80	52.08	2.65	54.73	good.	,,
,, (L 603 g) ....4.0 ft.	1.323	8.48	37.52	51.57	2.43	none.	...	

ULTIMATE ANALYSES, PERCENTAGE RESULTS.

	Moisture.	Carbon.	Hydrogen.	Oxygen and Nitrogen.	Sulphur.	Mineral Matter.	Combined Water.	Free Hydrogen.
Horunui Coal (L 602 a)..........	5.194	72.982	5.300	13.841	0.353	2.330	14.221	3.720
,, ,, (L 602 g)..	8.479	68.842	4.771	15.180	0.472	2.256	15.727	3.024

The outcrops of the beds as shown in detail on the map conform in a measure to the shape of the rock saddle ; though affected somewhat by the shape of the ground, especially on the north-western side of the saddle and near its summit where the dips are comparatively gentle. In fact the two or three lower beds are chiefly exposed in those parts of the survey by the cutting down of the larger of the several small valleys into the higher part of the arch of the saddle, so as to produce an oval shaped outcrop with the stream crossing it lengthwise. The same is true of the next two higher beds where they cross the still deeper valley of the larger branch above Haya-kawa village ; and the outcrops of those and the upper beds are still further complicated there by the small faults already mentioned. The outcrops of the two upper beds however sweep round the axis towards the north-east corner of the survey, and with the very steep dips of the south-eastern side of the saddle run almost straight to the south-west, but little affected by the variations of the surface of the ground

By means of the map the extent of the workable beds above certain levels has been measured ; and, taking into account the thickness of the beds and the steepness of the dip, the cubic contents have been calculated ; and the weight also by reckoning a cubic yard to the ton. The No. I Coal was found to underlie 97 acres (117,000 tsubo) above its lowest natural drainage level within the survey (about 685 feet above the sea) and to contain there (at a thickness of 3.7 feet) 590,000 tons. In like manner, within 500 feet below that level there are 161 acres (195,000 tsubo) or 1,030,000 tons ; making in both portions together 258 acres (312,000 tsubo) or 1,620,000 tons. Below the 500 foot level and less than 4,000 feet below sea level there are probably 1,064 acres (1,288,000 tsubo) or 9,505,000 tons ; making the whole amount of workable coal in the bed within the tract to be 1,322 acres (1,600,000 tsubo) or 11,125,000 tons.

Of the No. II Bed there are 188 acres (227,000 tsubo) or (at a thickness of 3.6 feet) 1,245,000 tons above water level (about 620 feet above the sea) ; and within 500 feet below that, 151 acres (183,000 tsubo) or 1,340,000 tons ; making together 339 acres (410,000 tsubo) or 2,585,000 tons. In addition, below the lower level and less than 4,000 feet below sea level there are probably 1,012 acres (1,225,000 tsubo) or 9,190,000 tons making in all 1,351 acres (1,635,000 tsubo) or 11,775,000 tons of workable coal.

There are likewise of the No. III Coal 159 acres (192,000 tsubo) or at a thickness of 3.3 feet 935,000 tons above water level (about 490 feet above the sea) ; and 130 acres (157,000 tsubo) or 1,060,000 tons within 500 feet below that ; making in both parts together 289 acres (349,000 tsubo) or 1,995,000 tons. Adding 1,029 acres (1,245,000 tsubo) or 8,615,000 tons that there are probably between the 500 foot level and 4,000 feet below sea level, we have in all 1,318 acres (1,594,000 tsubo) or 10,610,000 tons of workable coal of this bed within the survey.

The No. IV Coal has in the same way 97 acres (117,000 tsubo) or (at a thickness of 5.8 feet) 1,265,000 tons of coal above water level (about 420 feet above the sea) ; and within 500 feet below that, 112 acres (136,000 tsubo) or 1,655,000 tons ; making together 209 acres (253,000 tsubo) or 2,920,000 tons Besides, there are probably between that lower level and 4,000 feet below sea level 989 acres (1,197,000 tsubo) or 14,580,000 tons ; making in all 1,198 acres (1,450,000 tsubo) or 17,500,000 tons.

Of the No. V Coal there are above water level (about 415 feet above the sea) 98 acres (119,000 tsubo) or (at a thickness of 5.1 feet) 1,120,000 tons ; and within 500 feet of greater depth, 91 acres (110,000 tsubo) or 1,085,000 tons ; making together 189 acres (229,000 tsubo) or 2,205,000 tons. Moreover between the 500 foot level and 4,000 feet below sea level there are probably 971 acres (1,175,000 tsubo) or 12,550,000 tons; making all together 1,160 acres (1,404,000 tsubo) or 14,755,000 tons.

The No. VI Bed has above water level (about 360 feet above the sea) 56 acres (68,000 tsubo) or (at a thickness of 4.3 feet) 565,000 tons; and within 500 feet deeper 145 acres (175,000 tsubo) or 1,545,000 tons; making in both parts 201 acres (243,000 tsubo) or 2,110,000 tons. There are besides probably between the 500 foot level and 4,000 feet below sea level 800 acres (968,000 tsubo) or 8,480,000 tons; making in all 1,001 acres (1,211,000 tsubo) or 10,590,000 tons.

Of the No. VII Coal there are 72 acres (87,000 tsubo) or (at a thickness of 4.7 feet) 800,000 tons above water level (about 340 feet above the sea) ; and within 500 feet below drainage, 153 acres (185,000 tsubo) or 1,630,000 tons ; making both together, 225 acres (272,000 tsubo) or 2,430,000 tons. In addition there are probably between the 500 foot level and 4,000 feet below sea level 768 acres (929,000 tsubo) or '8,790,000 tons ; making in all 993 acres (1,201,000 tsubo) or 11,220,000 tons.

In all seven beds, then, within the tract, there are above water level 6,520,000 tons, and within 500 feet below that 9,345,000 tons ; making together 15,865,000 tons. Besides there are probably between the 500 foot levels and 4,000 feet below sea level 71,710,000 tons ; making in all 87,575,000 tons.

There are moreover at a less depth than 4,000 feet below sea level, though nowhere exposed within the survey certain portions of the Upper Bibai (five foot), Lower Bibai (three foot) and Sankebibai (four foot) Coal Beds, which may be reckoned in the workable coal of the tract. Of the Upper Bibai there are probably 884 acres (1,070,000 tsubo) or 11,450,000 tons of such coal ; of the Sankebibai, 72 acres (87,000 tsubo) or 830,000 tons ; and of the Lower Bibai 53 acres (64,000 tsubo), or 460,000 tons ; making in all three, 12,740,000 tons. Adding that to the amount of the seven exposed beds the whole amount of workable coal within the limits of the survey is found to be 100,315,000 tons.

5. *Shipment.*—Until last year there was no trace of a road within the limits of the survey, except a foot path from Poronaibuto to Hayakawa village ; but last summer Mr. Takabatake cut a line intended to become a horse road between those points (about 2½ miles or 1 ri), and later I ran and staked out a line for a steam railroad from the two exposures of the No. VI Coal Bed near Hayakawa village past Poronaibuto to the Ishcari River at Horumuibuto. The advantages and disadvantages of the line have already been so fully discussed in my report upon last season's work already in print that it is needless to go into the matter here so fully again. It is enough to say that the line is quite feasible, that it is 17½ miles (7 ri) long, and that it is the shortest line to water navigable for large vessels (with a draught of 10 or 12 feet), unless it be possible to build a road straight across the great swamp from two miles (⅘ ri) below Poronaibuto towards Bibaibuto or Bibaidap. Such a straight line would be about thirteen miles (5¼ ri) long from the mine to the Ishcari River ; and would be part of a very direct line only 17 or 18 miles (7 ri) longer to the town of Ishcari at the mouth of the River ; a line which, if built, would enable coal to be shipped all the year round instead of merely in the eight unfrozen

months, a very great advantage. The Ishcari is said to be at least ten feet deep in ordinary low water up to the mouth of the Bibai ; but the additional distance to be navigated from Horumuibuto is about 13 miles (5¼ ri).

The Ikushibets can at present scarcely be called a navigable stream for even log canoes from Poronaibuto downwards, and could hardly be converted into one for coal barges, or into a canal, without too great expense on account of the fall of about 100 feet to Horumuibuto which would require several locks. For the same reason, if for no others, a canal could hardly be dug across the swamp to Bibaibuto.

A horse road could be built from Poronai to Horumui- buto that would be somewhat shorter than the railroad line (which is everywhere a down grade line) ; but a horse road would be useless for shipping coal, unless provided with at least wooden rails ; and then it would be best to have it down grade all the way and follow nearly the course of the steam line. As, indeed, the steam power would become cheaper for a larger shipment than 300,000 tons a year it would probably be best to build any horse railroad in the place afterwards to be used for steam.

The feasibility of a railroad line is increased by the fact that it would tend to open up not only the remarkably rich Poronai field but many neighboring ones, the Ichiki- shiri field, the Nuppaomanai, and much more that has not yet been surveyed higher up the Ikushibets valley.

6. *Maps.*—The surveying of the coal field was begun in 1873, continued in 1874, and slightly added to in 1875, in connection with the railroad survey which was wholly made last year. In 1873, I was in the field here myself, aided by Messrs. T. Yamauchi, T. Kuwada, J. Takahashi and T. Saito, Assistant Geologists. In 1874, I made a visit of only a few days here, and left instructions as to

the work with the Assistant Geologists who did it, namely Messrs. T. Yamauchi, T. Kuwada, T. Kada, I. Ban,
E. Yamagiwa and S. Nishiyama. Last year I was aided
by Messrs. T. Inagaki, J. Takahashi, I. Ban, J. Shimada
and E. Yamagiwa, Assistant Geologists. Each surveyor
mapped his own work, and the whole was united into one
map (under my guidance of course) by Mr. Yamagiwa,
who with Mr. Takahashi in like manner did most of the
work of the final copy. The surveying was done with
prismatic compasses and pacing, except the railroad line
and some work near it, which were done with transit and
chain and served for the correction of the other work.
Stations two hundred feet apart (except on the railroad
line, where they were only 100 feet apart) were staked
and levelled with hand levels ; and, as shown on the
map, marked so as easily to be distinguished on the
ground. The geological observations were made either
by myself, or in my absence by Mr. Yamauchi. The
geological mapping and sections were chiefly done by Mr.
Ban under my oversight. The map of the coal field
(finished and dated 10th May, 1876) is on a scale of $\frac{1}{5000}$
and contains also a number of cross sections, to show the
dip and geological structure, and columnar sections to
show the bedding ; and the whole covers a space of 2.31
feet by 2.91 feet. It shows (like our other maps of the
same scale) by contour lines ten feet apart in level the
shape of the ground, and by broken contour lines of 100
feet apart in level the shape of the coal beds down to a
depth of 500 feet below water level. In order, however,
to prevent confusion, in the case of so many beds over-
lying one another, such lines have not been drawn for all
the beds on the final map ; yet the water level of each bed
is marked by a double line. The details of the map are
in general the same as in our other maps already described in their reports.

The railroad map (finished and dated 18 April, 1876,) was drawn (after I had myself plotted the railroad line) by Messrs. Takahashi and Yamagiwa. It is 2.70 feet high by 15.66 feet long, and is drawn in the same manner and on the same scale of $\frac{1}{8000}$; but only represents the surface of the ground by ten foot contour lines and the objects upon it, without attempting to display the underground geology, for which there were but two or three geological observations. Indeed the whole map (except small portions that are repeated in the Poronai or Nuppaomanai coal maps) lies quite outside the productive coal field, so far at least as the surface rocks are concerned and the greater part is a swampy plain of new alluvium. Not only is the railroad line and its immediate neighborhood given on the large map; but the whole course of the Ikushibets River below Poronaibnto. When distant from the railroad surveying the river is taken from sketches made by myself in canoe travelling, in 1873 and 1874; corrected of course here and there by the railroad survey. Upon the map it will be easy to lay down the course of such horse or waggon roads as may be desired.

A profile section of the railroad line is given in a separate drawing (by Mr. Takahashi) 0.88 feet high by 18.59 long, finished and dated 24th December, 1875. The section lengthwise is on the same scale as the map, $\frac{1}{8000}$; and vertically is drawn double, once on that same scale and once on a scale ten times exaggerated, which may be more convenient for engineering purposes. The levelling was only done with a hand level (by Mr. Takahashi); but an exploratory line and the final one were levelled separately and very carefully revised, and the whole length broken up into many polygons; so that the errors were not very large. Still it will be necessary, of course, before building the road to level it again with a more exact instrument, as we unfortunately could not do;

and then the grades of the line can be finally determined. In the mean time the section and map show very closely what the facilities for building the road are.

I have the honor to be,
Sir,
Your most obedient servant,

BENJ. SMITH LYMAN,
Chief Geologist and Mining Engineer.

Miyôhôji, Echigo ;
September 3, 1876.

GEOLOGICAL SURVEY OF HOKKAIDO.

REPORT

ON THE

ICHIKISHIRI

COAL SURVEY OF 1873;

ACCOMPANIED BY A GEOLOGICAL AND

TOPOGRAPHICAL MAP;

BY

BENJAMIN SMITH LYMAN,

CHIEF GEOLOGIST AND MINING ENGINEER.

TOKEI:
PUBLISHED BY THE KAITAKUSHI,
1876.

GEOLOGICAL SURVEY OF HOKKAIDO.

REPORT OF THE ICHIKISHIRI COAL SURVEY OF 1873 ;
ACCOMPANIED BY A GEOLOGICAL AND TOPOGRAPH-
ICAL MAP ; BY BENJAMIN SMITH LYMAN, CHIEF
GEOLOGIST AND MINING ENGINEER.

1.—SITUATION.
2.—LAY OF THE LAND.
3.—GEOLOGY.
4.—COAL.
5.—SHIPMENT.
6.—MAP.

His Excellency K, KURODA,
  Kaitakuchokuwan.

SIR :

I have the honour to make you the following report on
the Ichikishiri Coal survey.

1. *Situation.*—The survey covers a space of about
2,500 feet in width, one half of it on either side of a straight
line running about two miles and a half (one ri) S. 52°
W. (magnetic) from the south-west corner of the Poronai
survey towards the Horumui River, which in fact it
reaches by a narrow prolongation (probably quite outside
the coal field, about a mile and a half long (⅗ ri) and 600
or 800 feet wide; and contains therefore in the wider
part 750 acres (906,000 tsubo), or 1.2 square miles. The
line was run to explore the limits of the coal field towards

the Horumui and in the wider part of the survey side lines of a thousand feet in length were run right and left at every thousand feet, and the principal streams were also run.

2. *Lay of the Land.*—The survey includes a large part of the head waters of the Ichikishiri Brook, chiefly two or three of its larger branches, which cross the main part of the survey nearly at right angles in the middle and at either end. The main (or middle) branch is about 400 feet above the sea where it leaves the survey and the other two about 500 feet. The hills on either side of the branches rise to a height of about 900 or 1,000 feet and are generally very steep. Along the bottom of the valleys even, there is scarcely a trace of flat land. The whole surface is covered with forest and in very many places with tall bamboo reeds (*Arundinaria*). There is no human habitation within many leagues.

3. *Geology.*—The structure of the rock beds so far as known is extremely simple; being that of a saddle running lengthwise of the main part of the survey. Indeed, the intention was to have the main line run along the summit of the saddle (which had been observed in Poronai); and the discrepancy appears not to have been very great. Very few good rock exposures and only two or three of coal beds were found in the survey, as a guide to the studying out the structure ; but there were in addition several observations of bits of coal, or the "blossom" of the outcrops of coal beds ; and two or three of these having been discovered in our old note books, after we had made out the structure to our satisfaction and had drawn the outcrops on the map, proved to be in just the right places to confirm the accuracy of our work. With so few exposures we have not undertaken to make an independent columnar section of the beds ; but have taken it for granted that such a

section would be essentially the same as that of the adjoining Poronai survey ; and I beg to refer you to the Poronai Report for its details.

In the narrow prolongation of the Ichikishiri survey, near the north-eastern end and on the Horumui river, and on the main branch of the Ichikishiri a few hundred feet also below the survey there are exposures of vertical (85° dip) soft sandy and shaly gray rocks wth a strike of about north and south, quite different from the neighboring coal bearing rocks. The coal (strictly brown coal) bearing rocks belong to what I have called the Horumui Group. At the time of the Poronai and Ichikishiri surveys, three years ago, I supposed the nearly vertical rocks (exposed also on the Ikushibets below Poronaibuto), from their steeper dip and consequent older look (a slight degree of metamorphism, you would say) as well as their lower level, to underlie the coal bearing rocks; but, the next year, on seeing the same vertical rocks on the Ishcari River suddenly change before my eyes to a nearly horizontal dip through a long space, where they contained fibrous liquite, I was satisfied that they belonged to the Toshibets Group of rocks. Their lower level near the Ichikishiri and Poronai may arise from greater softness which has allowed them to be worn away more rapidly ; and it is probable too that the great boss of the Poronai and Ichikishiri saddle of the coal bearing rocks may have existed as an island at the time when the Toshibets Group was first deposited around it on the bed of the sea. The already folded and greatly solidified condition of the coal rocks may have partly kept them from further folding when the rocks of the Toshibets Group were so pressed together sidewise as to cause sharp north and south folds with nearly vertical dips ; but some traces of the north and south folding are to be seen at Poronai, as already mentioned in the report on that field, and possibly like traces may be found in the Ichiki-shiri coal field when more thoroughly explored. Possibly, as at Poronai, small north and faults may have been produced here by that folding, adding somewhat to the trouble of coal mining ; but it is not likely that they will prove very serious.

4. *Coal.*—No mineral of any economical importance except coal has been found on the Ichikishiri ; and of that only the same seven workable beds as at Poronai seem to crop out within the limits of the survey. Indeed only two of them Nos. II and III have been fairly opened here, but in several other places the outcrops of coal beds have been observed. None of the two or three openings were quite satisfactory ; for even the best (one of those on the No. III bed) seems to be just at the top of the saddle, and therefore perhaps does not give a very good measurement of thickness. It seems however safe to reckon upon the same thickness for the different beds as that shown by the numerous exposures in Poronai close by ; and a like assumption may be made in regard to their quality.

The outcrops of the beds are comparatively straight on the south-eastern side of the great saddle ; for the dip is very steep ($60°$ or $80°$), and the course of the outcrop is little affected by the variations in the surface of the ground. But the dip is much more gentle (say $15°$ or $20°$) on the other side of the saddle and the course of the outcrop is determined in a very great measure by the inequalities of the surface. The valleys of the streams have there cut down through the parallel, comparatively level beds of coal, and the outcrops consequently run around the bordering hill with only a gentle descent towards the north-west ; as the map shows in full detail.

By means of the map the extent of the different coal beds above certain levels has been measured ; and multiplying that by the number of feet in thickness of each bed and taking the steepness of the dip into account the number of cubic yards or of tons has been calculated. The south-western limits of the coal are not exactly known but are assumed to be at the end of the broad part of the survey ; for they must be very near it.

The No. I (or Bottom) Coal of Poronai has been found in that way to underlie 218 acres (264,000 tsubo) above its lowest drainage level within the survey (which is about 410 feet above the sea) ; and to contain therefore, at a thickness of 3.7 feet of good coal, 1,490,000 tons ; and within 500 feet below that level to underlie 228 acres (276,000 tsubo) containing 1,850,000 tons ; making in both portions together 446 acres (540,000 tsubo) or 3,340,000 tons. There are, besides, between the 500 foot level and 4,000 feet below sea level 261 acres (316,000 tsubo), or 3,710,000 tons ; making the whole amount of workable coal of the bed within the tract to be 707 acres (856,000 tsubo) or 7,050,000 tons.

In like manner the Poronai No. II bed has here 277 acres (335,000 tsubo) or (at a thickness of 3.6 feet) 1,775,000 tons above water level (which is about 370 feet above the sea) ; and within 500 feet below that, 129 acres (156,000 tsubo) or 960,000 tons ; making in both together 406 acres (491,000 tsubo) or 2,735,000 tons. In addition there are probably between the 500 foot level and 4,000 feet below sea level 233 acres (282,000 tsubo) or 3,250,000 tons ; making in all 639 acres (773,000 tsubo) or 5,985,000 tons of workable coal in the bed within the survey.

The Poronai No. III Coal has likewise 187 acres (226,000 tsubo) or (at a thickness of 3.3 feet) 1,100,000 tons above water level (about 440 feet above the sea) ; and within 500 feet below that, 132 acres (160,000 tsubo) or 1,150,000 tons ; making together 319 acres (386,000 tsubo) or 2,250,000 tons. Between the

500 foot level and 4,000 feet below sea level there are probably 216 acres (261,000 tsubo) or 2,780,000 tons ; making in all 535 acres (647,000 tsubo) or 5,030,000 tons of workable coal.

Of the Poronai No. IV Coal, with a thickness of 5.8 feet, there are 110 acres (133,000 tsubo) or 1,650,000 tons above the lowest drainage level of the main branch of the Ichikishiri (about 580 feet above the sea) ; and within 500 feet lower, 125 acres (151,000 tsubo) or 2,085,000 tons ; making in both parts together 235 acres (284,000 tsubo) or 3,735,000 tons, Below the 500 foot level and not 4,000 feet below sea level there are probably 182 acres (220,000 tsubo) or 4,015,000 tons ; making in all 417 acres (504,000 tsubo or 7,750,000 tons of workable coal.

The Poronai No. V Coal Bed, with a thickness of 5.1 feet, has 108 acres (131,000 tsubo) or 1,000,000 tons within the tract above the drainage level of the main branch of the Ichikishiri (about 590 feet above the sea) ; and within 500 feet below that, 95 acres (115,000 tsubo) or 1,365,000 tons ; making in both 203 acres (246,000 tsubo) or 2,365,000 tons. Between the 500 foot level and 4,000 feet below sea level there are 171 acres (207,000 tsubo) or 3,360,000 tons ; making the whole amount of workable coal in the bed within the tract to be 374 acres (453,000 tsubo) or 5,725,000 tons.

The No. VI Coal of Poronai, with a thickness of 4.3 feet, has 21 acres (25,000 tsubo) or 285,000 tons above water level (about 640 feet above the sea) ; and within 500 feet below that, 63 acres (76,000 tsubo) or 1,020,000 tons ; making together 84 acres (101,000 tsubo) or 1,305,000 tons. Besides, there are between the 500 foot level and 4,000 feet below sea level 73 acres (89,000 tsubo) or 1,180,000 tons; making in all 157 acres (190,000 tsubo) or 2,485,000 tons of workable coal.

Of the Poronai No. VII (or Top) Coal, with a thickness of 4.7 feet, there are 18 acres (22,000 tsubo) or 260,000 tons above water level (about 660 feet above the sea); and within 500 feet below that, 46 acres (56,000 tsubo) or 805,000 tons; making in both parts 64 acres (78,000 tsubo) or 1,065,000 tons. There are, besides, between the 500 foot level and 4,000 feet below sea level 53 acres (64,000 tsubo) or 930,000 tons ; making in all 117 acres (142,000 tsubo) or 1,995,000 tons of workable coal in the bed within the limits of the survey.

The deepest portion of the above given amounts of coal is indeed much less than 4,000 feet below sea level, and is a part of the No. I Coal on the south-east side of the saddle, where it is only 3,200 feet below sea level. On the north-west side of the saddle, even that lowest bed scarcely reaches downward within the tract to 100 feet above sea level.

In all seven beds, then, there are 7,560,000 tons above water level; and within 500 feet lower 9,235,000 tons; making together 16,795,000 tons. Between the 500 foot levels and 3,200 feet below sea level there are 19,225,000 tons; or in all 36,020,000 tons.

Moreover there are three other workable beds so low down as not to crop out anywhere within the survey; but exposed at only a few miles' distance and probably of good thickness here also; namely, the Upper Bibai, the Sankebibai and the Lower Bibai Coal Beds. The Upper Bibai coal (5 feet thick) contains within the tract at a depth less than 4,000 feet below sea level 8,100,000 tons, the Sankebibai Bed, four feet thick, in like manner, 435,000 tons and the Lower Bibai Bed, three feet thick, 95,000 tons; making an addition of 8,630,000 tons; or in all ten beds together 44,,650,000 tons.

5. *Shipment.*—There is at present not even a foot path (unless a deer track here and there) within the limits of

the survey ; but there would be no difficulty in building a road, even a railroad, down the main valley of the Ichikishiri to its mouth and thence by the same line as from Poronai to the mouth of the Horumui. The distance from the Coal to the mouth of the Ichikishiri would probably be about 5½ miles, or 2¼ ri ; and from the coal to the mouth of the Horumui, about 17½ miles or 7 ri, the same as from the Poronai coal to that point. Whenever a railroad shall be built from Poronai it will undoubtedly be best to build also a branch line from the mouth of the Ichikishiri to the present survey ; and although the amount of coal opened up by such a branch will be less than the coal of Poronai, yet it will be large enough, and the distance will be about the same. For, even if the Poronai railroad should go across the mouth of the Bibai, it will have to pass near the mouth of the Ichikishiri.

6. *Map.*—The map was finished and dated the 13th of May, 1876. It is on a scale of $\frac{1}{5000}$, has two unexaggerated cross sections on the same scale, one lengthwise of the axis and one across it to show the structure, and a small map to show the general position of the field ; and the whole covers a space of 0.9 foot by 4.29 feet. The shape of the ground is shown by contour lines ten feet apart in level ; and the shape and position of the seven coal beds is shown by lines that represent the outcrop and the place of the lowest water level in the valley of the main branch of the Ichikishiri within the limits of the survey. Two of the beds, Nos. IV and V, would have a still lower water level on another branch to the south-west, but the extent of the coal so far in that direction is perhaps too doubtful to be counted on so certainly. The beds to be represented are so numerous within a small space that it seemed likely to be too confusing if hundred foot contour lines upon them should be drawn as has been done for some of the coal beds of the maps of other surveys.

The staked stations, 200 feet apart on the straight lines, and commonly less on the streams, are distinguished as on our other maps and surveys by separate marks so that any point may be found on the ground. The other details of the map are likewise in general the same as on our other maps

The survey was made in 1873 by myself with the aid of Messrs. T. Yamauchi, T. Kuwada, J. Takahashi and T. Saitô, assistant Geologists. The lines were run with prismatic compass and pacing; and the main line was levelled with a hand level, the side lines with aneroids, except one or two which failed to be levelled at all. The surveyors generally plotted each his own line, and the whole was united into one map by Mr. Kuwada. The geological observations were by myself; but the geological mapping and sections were studied out and drawn, with a little general instruction from me, by Mr. I. Ban, Assistant Geologist, after his similar work on the Póronai map, lately finished; and were so intelligently and carefully done as to need very slight corrections on my revision of them; a proof of the progress that our assistants have made in the art of geological surveying. Mr. Ban also had charge of making and lettering the finished copy, but was aided a little by some of his comrades, in our last days at Yedo.

I have the honor to be,

Sir,

Your most obedient servant,

BENJ. SMITH LYMAN,
Chief Geologist and Mining Engineer.

Miyôhôji, Echigo,
3rd September, 1876.

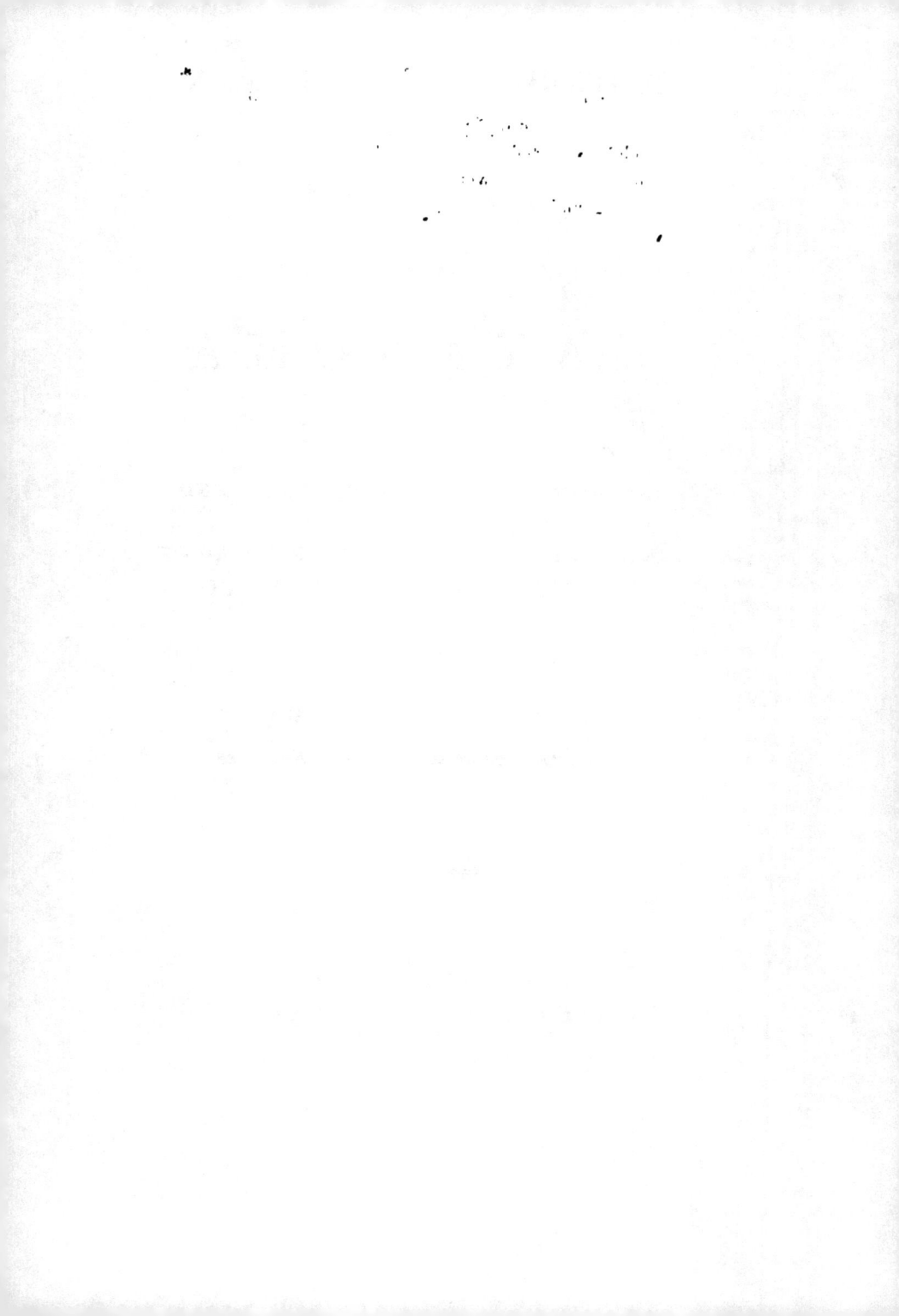

GEOLOGICAL SURVEY OF HOKKAIDO.

REPORT

ON THE

KAYANOMA

COAL FIELD;

ACCOMPANIED BY A GEOLOGICAL AND
TOPOGRAPHICAL MAP OF THE
FIELD AND BY A HYDROGRAPHIC MAP OF
THE PROPOSED HARBOR AT SHIBUI;

BY

BENJAMIN SMITH LYMAN,

CHIEF GEOLOGIST AND MINING ENGINEER.

TOKEI:
PUBLISHED BY THE KAITAKUSHI,
1876.

GEOLOGICAL SURVEY OF HOKKAIDO.

REPORT ON THE KAYANOMA COAL FIELD ; ACCOM-
PANIED BY A GEOLOGICAL AND TOPOGRAPHICAL
MAP OF THE FIELD AND BY A HYDROGRAPHIC
MAP OF THE PROPOSED HARBOR AT SHIBUI ; BY
BENJAMIN SMITH LYMAN, CHIEF GEOLOGIST AND
MINING ENGINEER.

1.—SITUATION.
2.—LAY OF THE LAND.
3.—GEOLOGY.
4.—COAL.
5.—SHIPMENT.
6.—MAPS.

His Excellency K. KURODA,
          Kaitakuchokuwan,

SIR :

I have the honor to make you the following report on
the Kayanoma Coal Field.

1. *Situation.*—The Coal Field lies upon the valleys of
the Tamagawa whose mouth is at Kayanoma village, about
seven miles and a half, or 3 ri, north of Iwanai, on the
west Coast of Yesso, and the adjacent Shibui and Chatsunai valleys on the south. The survey runs about three
miles (1½ ri) back from the shore and nearly as far long
it and covers 5.9 square miles; but probably only a small
part of that space has coal bearing rocks on its surface.

As the old volcanic, tufaceous rocks which limit the field seem to be nearly or quite conformable to the coal bearing rocks it is not unlikely that some of the coal beds may be found underneath the volcanic rocks, adding very much to the extent of the field. It is also very possible that the coal rocks reach to a short distance beyond the north-western edge of the survey; though probably not far.

2. *Lay of the Land.*—The valley of the Tamagawa with a south-westerly course fills the north-western side of the survey; and the smaller Chatsunai valley, nearly parallel, takes up the central portion, leaving a small space for the Shibui valley between it and the Tamagawa; and the Iwanai river with its large northerly branch, the Hattari, flows along the south-western edge and empties into the sea at Horikap village, the southernmost corner of the survey.

Although there is a wide tract of flat land to the south of the Iwanai and Hattari rivers and 300 or 400 yards in width on the northerly side for a mile or so from the sea, there is no more of level ground within the survey except a very little on the Tamagawa one or two hundred yards wide. The hills between the valleys rise gradually higher and higher from the sea shore until a height of about 600 feet is reached at the middle of the survey and one of 1,300 feet at the north-eastern end. In general the hills are very steep; but towards the sea shore somewhat more gently rounded.

The land at a little distance from the sea is still covered with wild forest and a thick tall growth of dwarf bamboo (*Arundinaria*); but on the shore there are several villages surrounded by farms and cleared land, and the clearings extend here and there to about two miles and a half (one ri) up the Tamagawa. It is there that the old mines of a dozen years or more ago (the Hurushiki mines, now abandoned) were worked; and it is half a mile nearer the sea where the present mines were opened about ten years ago, on a small branch of the Tamagawa. There are two or three houses there and across the hill in the Chatsunai valley near by ; and a small miners' village half a mile further down the Tamagawa valley ; then half a mile above the mouth of the Tamagawa there is another small village for men connected with the mines. There are also villages on the sea shore, Kayanoma at the mouth of Tamagawa, Shibui and Chatsunai at the mouths of streams of the same name, and Horikap at the mouth of Iwanai River, all small fishing villages ; besides a few scattered houses and hamlets in little nooks along the shore between. On the Hattari is a scattered farming settlement.

Near the present mines the main branch valley is the Osawa, with a fork called the Kosawa on the north-east next to the mines, and one called the Takarasawa between that and the main branch, and the Ogurasawa on the same side of the Osawa further up About 300 yards south-west of Osawa is the Hamba valley another nearly parallel small branch of the Tamagawa.

3. *Geology.*—The general geological structure of the coal bearing rocks seems to be very simple ; and to be that of a broad saddle plunging rapidly seaward, with its axis nearly in the bottom of the Chatsunai valley, running about N. 50° E., magnetic, or say north-east and south-west, true bearing ; and of a rather flat basin (also plunging seaward) on the north-west with its axis in the Kirarayama hills that border that side of the Tamagawa valley. There are however in addition several faults of small extent, of which four have been discovered, and probably others remain to be discovered and they seem to have a course of about North 30° West (magnetic). They are probably the result of some effort to push the already folded rock beds into new folds of that direction (as may

be seen in a like case at Poronai); and the effort may in some parts of the field have resulted in bending the beds out of their usual shape without breaking them into faults. The extent and dip of the faults already discovered are not yet known.

The dip of the rocks on the north-western side of the great saddle is generally about 50 degrees, as seen in the present mines ; but sometimes for a small space, owing to local disturbance near a fault it has become almost vertical, as on the Ogurasawa. On the south-easterly side of the saddle and of Chatsunai valley the dips are gentler, about 30 degrees ; and on the Kirarayama side of the northerly basin the dips are only about 15 degrees. On the seaward end of the main saddle the dips are also no doubt gentle towards the sea ; but no very good dip observations, none at all of coal, have yet been made there.

Besides great cliffs of old volcanic, tufaceous rocks on the sea shore and other exposures of old volcanic rocks on the outskirts of the survey, and a little new alluvium along the Tamagawa and the Hittara the only rocks within the survey are the coal bearing rocks and these alone have been chiefly studied. They belong without doubt to the same age as the brown coal bearing rocks of Poronai and the Ishcari valley, the Horumui Group and a certain resemblance to a part of the Poronai section may be seen in the following sections (from above downwards) which have been made out from a combination of all the observations near the Kayanoma mines by means of the geological and topographical map :—

Feet.

Lime rock (in Hamba valley)................................ 3.00
Hidden.................... ....245.00
Hard black slate, (L 337 g), Hamba No. 1.......... 4.50
Gray shaly sand rock.,................................... 1,50

	Feet.
Hidden..	44.00
Black slate, (L 338 h), Hamba No. 2 (probably the same as the Tamagawa No. 1)..........................	1.60
Light gray sandy shale.........	0.50
Hidden	8.50
Brownish black hard sandy slate.......................	0.50
Good *coal*, (L 326 t), Osawa No. 3, (probably same as the Tamagawa No. 2 and Onkosawa No. 6)............................1.20	
Yellow clay ....,..0.10	
Bony *coal* and slate.............................6.00—	7.30
Gray clay	0.10
Floor like roof, about.......................................	0.50
Hidden..............................	6.50
Whitish gray clay .... ~................	0.90
Dark brown sandy shales, roof	0.20
Bony *coal*..0.50	
Parting, same as roof..............................0.05	
Good *coal* (same as the Hurushiki, Hamba No. 3 and Oukosawa No. 5) (L 326 u), Osawa No. 4.,...............................1.25	
Dark clay...0.05	
Good *coal* 0.20	
Parting like roof............. .... 0.05	
Good *coal*..0.80	
Parting like roof.................................0.70	
Gray clay 0.10	
Coal, pretty good...........................1.00	
Parting like roof...... ....0.20	
Good *coal* 0.60	
Gray clay 0.10	
Bony *coal* 0.50	
Gray clay 0.03	
Good *coal* ....,.............................1,32—	7.45

	Feet.
Dark brownish black sandy shale, about	4.00
Hidden	56.00
Sand rock, light gray, shaly, soft	4.00
Roof, dark brown fine grained	0.50

Coal...0.20

Dark mud rock 0.06

Bony *coal* 1.14

Good *coal* 0.70

Parting, gray clay, hard..........................0.20

Middling *coal* 1.10

Parting 0.10

Middling *coal*, *(L* 324 r*)*, Osawa No. 1	
(same as the Midzunukishiki No. 1 and	
Onkosawa No. 4)..... ....1.20—	4.60
Black slate parting	0.10
Sand rock, light gray	1.20
Soft or shaly sand rock	30.00
Roof slate	0.50

Coal, Midzunukishiki No. 2...................2.40

Parting............................. . 0.30

Coal.....................................0.30—	3.00
Hidden	16.50
Soft dark brown shale..............................	0.50
Roof, hard black slate	0.50

Good *coal*, (L 327 a), Osawa No. 5 (same as
the Osawa No. 2, Kosawa No. 1, Honshiki, Hamba No. 4 and Onkosawa
No. 3)......................................1.25

Bony *coal*...................................0.85

Light yellow gray clay..........................0.30

Coal, good ....1.00—	3.40
Dark brown shale..............................	2.50
Slate and *coal* mixed	1.20
Hard brownish black slate	2.00

	Feet.
Soft dark brown shale	1.00
Soft or shaly sand rock	21.50
Light gray soft shale	1.50
Hard black slate	0.45
Slate and *coal* mixed	2.50
Light gray clay	0.70
Bony *coal* and slate	1.10
Coal	0.20
Blue gray clay	0.10
Black slate	0.45
Good *coal*	1.00
Gray clay	0.50
Good hard *coal*, (same as the Kosawa No. 2, Onkosawa No. 1, Kirarayama Shimonose, and Osawa No. 6)	2.90
Hard parting like roof	0.10
Coal	0.20
Hard parting like roof	0.10
Coal, bad,	0.70
Soft gray shaly clay	2.00
Coal	0.60
Parting like roof	0.20
Good *coal*	0.90— 9.95
Hard parting like roof	0.30
Soft brownish gray shale	0.70
Hidden, probably soft or shaly sand rock	1.80
Coal and slate mixed	0.50
Soft brownish gray shaly clay	1.20
Coal and slate mixed	2.00
Soft gray clay shale	1.10
Slate and *coal* mixed	1.10
Soft gray shaly clay	0.90
Bad *coal*, Osawa No. 7 (same as the Kosawa No. 3 and the lower part of Onkosawa No. 1 and of Kirarayama Shimonose)	2.50

	Feet.
Soft gray brownish clay, exposed......................	2.00
Soft or shaly sand rock...............	18.50
Soft gray shale......................	0.50
Roof, hard black slate...................	0.55
Bony *coal* and slate.................	0.75
Coal 0.40	
Parting, hard like roof.......................... 0.05	
Good *coal*, Osawa No. 8 (same as the Taka-	
rasawa No. 2, Kosawa No. 4, Tateire	
No. 2, and Chatsunai Shimonosawa	
No. 2)......................................1.40	
Hard parting, like roof.............................0.15	
Coal ....0.30—	2.30
Brownish black sandy rock	2.00
Soft yellow shaly clay	2.30
Coal 0.80	
Yellow clay 0.05	
Good *coal* ....,............... ....1.40	
Yellow parting 0.05	
Good *coal*..................... ...,.,............1.20	
Mixed shale and bone 0.60	
Yellow clay 0.10	
Good *coal*, (L 329 e), Osawa No. 9 (same	
as the Takarasawa No. 3, Hamba No.	
5 and Chatsunai Shimonosawa No. 1) 1.80	
Bluish gray clay 0.15	
Good *coal* 1.40	
Bluish gray clay 0.05	
Bad *coal* 0.30	
Parting, hard like roof............................0.60	
Good *coal*, hard ....1.50—	10.00
Floor, black slate and *coal* mixed......................	1.30
Soft or shaly sand rock	50.00

	Feet.
Roof, brownish black slate	1.30
Coal, bad, (L 424 h) Ogurasawa No. 1 (same as the Takarasawa No. 4, Kosawa No. 5, Tateire No. 3 and Kiyosumisawa No. 2)	0.50
Black hard slate	0.30
Soft dark gray shale	0.50
Brown to gray clay.	2.00
Hidden	19.50
Roof, hard brown black slate	0.80
Good coal (L 424 i) Ogurasawa No. 2 (same as the Kiyosumisawa No. 1)	2.60
Floor, like roof	0.80
Hidden	36.50
Roof, black hard shale	0.60
Bony coal	1.70
Soft brown clay	0.70
Bony coal	0.50
Pretty good coal	2.00
Gray and brown clay	0.30
Pretty good coal, (L 424 j), Ogurasawa No. 3.	4.00
Gray shale, soft	1.00
Hard black coal slate	0.40
Pretty good coal	1.40— 12.00
Gray shale	0.50

681.45

Many of the coal beds of the main section have also been measured at several other points ; and the following are the results for the principal beds (in all cases from above downwards) :

The Tamagawa No. 3 or Hurushiki bed (probably the same as the Osawa No. 4) gave :

Dark brownish black shale	1.20
Coal slate and coal mixed	2.00

	Feet.
Light gray shale, rather hard	6.00
Coal, exposed 2 ft. or more, and said to be in all...	6.00
	15.20

The same bed at Hamba No. 3 gave :

Black hard sandy slate	0.30
Coal	1.00
Mixed slate and *coal*	0.50
Black hard sandy slate	0.25
Soft light gray sandy shales	3.00
	5.05

The same bed again at Onkosawa No. 5, (L 417 ga), gave :

Hard brownish black slate	0.30
Good *coal*	2.00
Coal and slate mixed	2.20
Hard brownish black slate	0.30
	4.80

The Midzunukishiki (No. 1) bed (the same as the Osawa No. 1) gave :

Coal with a very thin parting of slate in the middle	4.00

The same bed at Onkosawa No. 4 gave :

Roof, soft sandy shales, exposed about	20.00
Black slates with 0.1 foot of *coal* in the middle	6.50
Floor like roof,	4.00
	30.50

The Honshiki bed, that of the principal mine, (the same bed as Osawa No. 5) gave in one part of the mine :

Coal	6.40
Slaty parting	0.30
Coal	1.30
	8.00

Feet.

In another part of the mine it gave :

Coal 4.20

But perhaps that was only the upper bench.

The same bed at Kosawa No. 1, (L 332 b), gave :

		Feet.
Brownish shale...............................		0.60
Bad and bony *coal* and slate mixed		1.10
Good *coal*, soft 2.65		
Good *coal*, rather harder ....5.80—		8.45
Hard dark parting		2.00
Soft gray shale...		4.00
Mixed slate and bony *coal*		2.30
Soft dark gray shale..		1.00

19.45

The same bed again at Osawa No. 2, (L 325 s), gave :

Light gray sandy clay..............................		0.50
Dark brownish black sandy shale		0.35
Bony *coal* 0.80		
Good *coal*...............................1.00		
Brownish black sandy slate..................0.10		
Good *coal*...............................3.90		
Bony *coal* 0.35		
Very bony *coal*...........................2.60—		8.75
Dark brown sandy mud rock...........................		0.80
Bluish gray and yellow clay............................		1.20
Bad *coal*		0.50
Dark brown hard mud rock............................		1.90
Good *coal*..		0.50
Dark brown mud rock..................................		0 20
Soft light gray shale.................................		1.00

15.70

Feet.

The same bed again at Hamba No. 4, (L 340 j), gave :

Brown and dark gray clay..............................	1.00
Brownish black sandy slate...........................	0.60
Coal............................. ....0.30	
Black slate..2.10	
Coal...0.35—	2.75
Hard black sandy shale.................................	0.80
Soft light brown shaly clay............................	2.00
Black slate..	3.00
Hard brownish black sandy shale.....................	0.50
Soft dark gray clay	0.50
	11.15

The same bed again at Onkosawa No. 3 gave :

Brownish black shale	0.50
Good coal	2.00
Soft brownish black shales	0.60
	3.10

The Onkosawa No. 1, or (L 417 i), bed (the same in the upper part probably as the Osawa No. 6 and in the lower part as Osawa No. 7) gave :

Brownish black shale	0.30
Coal 5,00	
Gray soft clay 0.10	
Coal ....,...........0.70	
Sandy shale 0.35	
Good coal 1.40	
Sandy shale 0.35	
Mixed slate and coal..............................1.00	
Bad bony coal 1.00	
Good coal 0.90	
Black hard parting 0.10	
Good coal ....0.65—	11.55

	Feet.
Mixed black shale and *coal*....................................	1.90
Bluish gray sandy shale, soft...........................	0.40
Bad *coal*..	1.30
Blue sandy shales..	0.30
Black parting...	0.30
	16.05

The Kosawa No. 2 (the same also as Osawa No. 6) gave :
Hard black shaly fine grained sand rock with thin

layers of *coal*...		2.30
Coal (0.7 good, the rest poor)................1.90		
Clay 0.02		
Good *coal*...0.25		
Parting..0.10		
Poor *coal*...0.30—	2.57	
Hard brownish black sandy slate......................		0.50
Gray clay, exposed about...............................		1.50
		6.87

The Takarusawa No. 3 bed (the same as the Osawa No. 9) gave :

Gray soft shale...		1.00
Good *coal*............................... ....1.40		
Hard black slate parting...........................0.20		
Good *coal*...6.00		
Black hard slate......... ....0.30		
Good *coal*........ ....1.20		
Black hard slate......................................2.00		
Good *coal*...0.50—	11.60	
Hard black slate..............................		1.80
		14.40

Feet.

The same bed at Hamba No. 5, (L. 340 j), gave:

Coal... 1.50

Gray hard shales, exposed about....................... 3.00

_____

4.50

The same bed again at Chatsunai, Shimonosawa No. 1, (L 414 a), gave:

Roof, soft black shale.................................... 1.50

Shale and coal mixed.................................... 6.20

Floor like roof............. .... 0.30

_____

8.00

The correspondence of the beds is, however, not so certain in this survey as it has been in those of the Ishcari valley.

4. *Coal.*—The coal (strictly speaking, brown coal) is the chief, if not the only, mineral of economical importance within the limits of the survey. It may be seen in the sections that six of the beds have a thicknes of more than three feet and may therefore be considered workable; but three of them, the Hurushiki, the Honshiki and the Onkosawa No. 1 beds are decidedly better than the other three, the Midzunukishiki, the Takarasawa No. 3, and the Ogarasawa No. 3.

As may be seen by comparing the sections, all the beds appear to vary greatly in thickness; so that they are generally not workable throughout their whole extent. The Onkosawa bed however seems to have a good thickness at every exposure but one, and that may be near a fault and therefore a very local thinning out of the bed which could not appreciably affect the whole amount of the coal.

The quality of several of the coals has been tested by chemical assays and the results are given in Mr. Munroe's report of 1874 on "Yesso Coals." The true relative

position of the different beds had at that time been far
less satisfactorily ascertained than it has been now ; so
that a few words are necessary here to explain to which
of the above mentioned beds the assayed specimens belonged ; and some of the principal results of the assays
and analyses may be recalled by the two accompanying
tables.

In descending order :

The Hurushiki Coal is probably the same as the Osawa
No. 4, the Hamba No. 3 and the Onkosawa No. 5.

The Midzunuki Coal is the same as the Osawa No. 1
and probably the Onkosawa No. 4.

The Honshiki Coal is the same as the Kosawa, No. 1,
the Osawa No. 2, Osawa No. 5 and probably the Hamba
No. 4 and the Onkosawa No. 3.

The Osawa No. 6 coal is the same as the Kosawa No.
2, and probably the Onkosawa No. 1 and the upper part
of the Kirarayama Shimonose.

The Tateire No. 2 coal is the same as the Kosawa No.
4 and probably the Takarasawa No. 2, the Osawa No. 8
and the Chatsunai Shimonosawa No. 2.

The Osawa No. 9 coal is the same as the Takarasawa
No. 3, and probably the Hamba No. 5.

The Ogurasawa No. 3 seems to be exposed nowhere
else.

ASSAYS.

Name of Coal.	Specific Gravity.	Moisture at 140° C.	Volatile Combustible.	Fixed Carbon.	Ash.	Per centage of coke.	Appearance of coke.	Color of Ash.
Hurushiki	1.334	1.34	28.44	54.53	15.69	70.22	excellent.	Lavender.
Midannuki coal average sample....4.0 ft.	1.446	3.73	33.03	41.76	21.48	63.78	good.	Dk. reddish brown.
Honshiki coal upper bench....6.4 ft.	1.266	5.71	39.94	50.63	3.72	54.35	very good.	Light brown.
" " lower bench....1.3 ft.	1.340	5.51	33.39	46.03	14.07	60.10	fair.	Light lavender.
" " average sample from mine...	1.351	5.36	35.95	46.11	13.08	59.19	very good.	Light brown.
" " from "dump".	1.363	4.12	34.88	45.26	15.74	61.00	very good.	Light brown.
Osawa No. 6, upper benches....3.9 ft.	1.286	5.43	41.43	43.36	9.78	53.14	good	Reddish brown.
" ", lowest (?) benches....1.5 ft.	1.589	4.85	25.30	29.79	40.06	none.	......	White.
Tateire No. 2, upper bench....1.9 ft.	1.355	5.63	32.22	49.38	12.77	none.	......	Reddish brown.
" " middle bench....1.1 ft.	1.539	3.43	25.63	36.06	34.86	none.	......	White.
" " lower bench....0.7 ft.	1.346	5.66	33.68	48.71	11.95	60.66	fair.	White.
" " average sample from mine	1.411	5.06	29.93	41.50	23.51	65.01	very poor.	Brownish white.
Kosawa No. 4, upper bench....1.6 ft.	1.359	12.98	41.84	38.05	7.13	none.	......	Pink.
" " middle bench....1.7 ft.	1.367	12.71	44.97	36.55	5.77	none.	......	Pink.
" " lower bench....1.0 ft.	1.408	6.19	31.75	44.42	17.64	none.	......	Reddish brown.
" " calculated average of bed	1.372	11.51	41.07	38.73	8.69	none.	......	Light brown.
Chaïsunai, Onkosawa No. 1; 5 ft. bench	1.334	10.85	39.63	43.46	6.06	none.	......	Reddish brown.
Osawa No. 9, 6th bench....1.4 ft.	1.347	6.75	39.03	42.57	11.65	54.22	fair.	Light brown.
" " 5th bench....1.2 ft.	1.301	5.66	41.47	46.44	6.43	52.87	good.	Light brown.

ASSAYS.—*Continued*

Name of Coal.	Specific Gravity.	Moisture at 140° C.	Volatile Combustible.	Fixed Carbon.	Ash.	Per centage of coke.	Appearance of coke.	Color of Ash.
Osawa No. 9, 4th bench..........1.8 ft.	1.296	9.40	38.66	46.47	5.47	none.	......	Light ochre.
" " 3rd bench..........1.4 ft.	1.316	9.55	35.80	49.71	4.94	none.	......	Pink.
" " calculated average of bed..	1.315	8.03	38.60	46.20	7.08	??	very poor?	Light brown?
Ogurasawa No. 3, upper bench......2.0 ft.	1.400	6.43	35.33	38.44	19.80	none.	......	White.
" " middle bench......4.0 ft.	1.410	7.86	35.03	39.28	17.83	none.	......	White.
" " lower bench......1.4 ft.	1.423	5.91	36.27	33.98	23.84	none.	......	White,
" " calculated average	1.410	7.04	35.37	37.97	19.62	none.	......	White,

ULTIMATE ANALYSES, PER CENTAGE RESULTS.

	Moisture.	Carbon.	Hydrogen.	Oxygen and Nitrogen.	Sulphur.	Mineral matter.	Combined water.	Free Hydrogen.
Hurushiki coal	1.342	69.049	5.256	7.172	2.386	14.795	6.718	4.510
Midzunuki coal (mine av.) ...	3.714	57.689	4.620	10.144	3.765	20.068	10.062	3.502
Honshiki coal (mine av.)	5.350	65.221	5.222	10.118	1.607	12.472	10.033	4.107
Honshiki coal (dump av.)	4.095	64.412	4.911	9.940	1.449	15.193	9.832	3.818
Tateire coal (mine av.)	5.060	56.283	4.124	10.271	1.178	23.084	10.205	2.990

It appears therefore that the Ogurasawa No. 3 Coal, although of such ample thickness, is of comparatively poor quality owing to its large amount of ashes. It is possible however that it may yet be thought workable, especially if it should be found that a good portion of the stony matter could be separated by careful mining or by crushing and washing.

The Hurushiki bed, although its working was abandoned on account of its greater distance from the sea has the peculiarity of coking very well indeed unlike all the other Yesso coals yet examined ; and is therefore very valuable for iron making, and deserves to be reopened.

The outcrops of the different beds are in general nearly parallel to each other, and conform to the shape of the basin and saddle ; retreating far from the sea in the Kirarayama basin, approaching the shore on the Chatsunai saddle, and running back inland on the southerly side of the saddle, past the exposure of Onkosawa. The details are more fully shown on the map of the coal field.

The same map has enabled the extent of certain portions of the different coal beds to be measured ; and taking account of their thickness and of the steepness of their dip, the cubic contents have been calculated and the weight in tons, reckoning a cubic yard to the ton. In order not to exaggerate the amount of workable coal only small portions of each bed, those that may be most safely considered to have a permanently workable thickness, have been measured ; but the Onkosawa No. 1 on account of its less variable thickness, has been measured through a wide extent. It is not unlikely, however, that the beds may prove in fact to be workable through a much wider space (especially if some of the thin measurements on the Osawa and in Hamba valley be caused by the nearness of faults and therefore be only local) ; or that if they do really thin out so extensively, their places may be made good by other beds that have become thicker in the same portions of the field. The measured amounts of workable coal then are probably far from being over-statements of what really exists within the tract.

The Hurushiki bed, then, was measured only from the north-western edge of the survey (on Kirarayama) to the easternmost of the four faults which traverses the pre-sent workings of the Honshiki mine, in fact only through a small corner of the whole survey ; and was found to have there 40 acres (48,000 tsubo) above its lowest na-tural drainage level, about 240 feet above the sea ; or, at an average thickness of six feet, 250,000 tons. Within 500 feet below that level in the same space there are 55 acres (67,000 tsubo), or 650,000 tons ; making in both portions of the bed together 95 acres (115,000 tsubo), or 900,000 tons. Below the 500 foot level, reaching at the edge of the tract to not more than 3,000 feet below sea level, there are 35 acres (42,000 tsubo), or 400,000 tons ; making the whole workable amount to be 130 acres (157,000 tsubo), or 1,300,000 tons.

The Honshiki bed, in like manner, was only measured through a small corner of the whole survey, namely from the north-western edge of the survey to the westernmost fault, the one that passes along the Osawa valley, and found to have within that space, above the low water level of the exposure Osawa No. 2, about 290 feet above the sea, 35 acres (42,000 tsubo) or, at a thickness of six feet, 400,000 tons; and within 500 feet below that level, 75 acres (91,000 tsubo) or 1,000,000 tons; making, both to-gether, 110 acres (133,000 tsubo) or 1,400,000 tons. Below the 500 foot level and not more than 3,000 feet below sea level even at the edge of the tract, there are 130 acres (157,000 tsubo) or 1,600,000 tons ; making in all 240 acres (290,000 tsubo) or 3,000,000 tons.

The Onkosawa No. 1 bed was measured from the

north-western edge of the survey all round (except un-
surveyed land) to a line half way between Onkosawa and
the eastern edge, and extending at the greatest workable
depth even under the old volcanic rocks (which are
probably not so thick as that) to the sea shore and a few
score yards beyond. It was found that there were within
that space 70 acres (85,000 tsubo) or (at an average
thickness of six feet) 1,500,000 tons above water level
(about 240 feet above the sea) ; and within 500 feet
below that level, 200 acres (242,000 tsubo) or 2,500,000
tons ; making, together, 270 acres (327,000 tsubo) or
4,000,000 tons. Between the 500 foot level and 4,000 feet
below sea level there are probably 1,260 acres (1,525,000
tsubo) or 20,900,000 tons ; making the whole amount to
be 1,530 acres (1,852,000 tsubo) or 24,900,000 tons.

These three best beds together then may be counted on
as containing 2,150,000 tons of workable coal above water
level and within 500 feet below that, 4,150,000 tons ; or,
in both parts together, 6,300,000 tons. In addition there
are between the 500 foot levels and 4,000 feet below sea
level probably 22,900,000 tons ; making in all 29,200,000
tons.

But the Midzunukishiki bed has also been estimated
(by comparison with the others) from the north-western
edge of the survey to a line drawn half way between the
Osawa and Hamba valleys, and found to have in that
space probably at least 10 acres (12,000 tsubo) of workable
coal, or, at an average thickness of $3\frac{1}{2}$ feet. 90,000 tons,
above the low drainage level of the Osawa No. 1 exposure
(about 250 feet above the sea); and within 500 feet below
that level, 36 acres (44,000 tsubo) or 200,000 tons; making, both together, 46 acres (56,000 tsubo) or 290,000
tons. Besides that, there are probably between the 500
foot level and at most 3,000 feet below sea level (to the
edge of the survey) 145 acres (175,000 tsubo) or 1,200,000

tons; making in all 191 acres (231,000 tsubo) or 1,500,000 tons.

The Takarasawa No. 3 bed in like manner within the same space contains probably at least 10 acres (12,000 tsubo) or (at a thickness of seven feet) 180,000 tons above water level (say 260 feet above the sea); and within 500 feet below that level 36 acres (44,000 tsubo) or 400,000 tons; making, both together, 46 acres (56,000 tsubo) or 580,000 tons. There are besides between the 500 foot level and the edge of the survey (atmost 3,000 feet below sea level) probably 150 acres (181,000 tsubo) or 26,000,000 tons; making in all 196 acres (237,000 tsubo) or 3,200,000 tons.

The Ogurasawa No. 3 bed has likewise been estimated for the same space as the Honshiki was measured, namely from the north-western edge of the map to the western-most fault, and found to have there probably 50 acres (60,000 tsubo) or (at a thickness of 7½ feet) 750,000 tons above water level (say 280 feet above the sea); and within 500 feet below that level, 85 acres (103,000 tsubo) or 1,300,000 tons; making in both parts together 135 acres (163,000 tsubo) or 2,050,000 tons. In addition there are between the 500 foot level and the edge of the survey (at most 3,000 feet below the sea) probably 170 acres (206,000 tsubo) or 2,600,000 tons; making in the whole 305 acres (369,000 tsubo) or 4,650,000 tons.

In all six beds, then, there are probably at least 3,200,000 tons above water level, and 6,050,000 tons within 500 feet below that; making together 9,250,000 tons. Moreover between the 500 foot levels and 4,000 feet below sea level there are probably 29,300,000 tons ; making the whole amount of workable coal within the limits of the survey to be at least 38,550,000 tons.

Of that amount only an insignificant portion has so far been worked out or made unworkable ; probably little

more than two acres (2,500 tsubo) or say 100,000 tons,
of which perhaps only a small portion has been made
available for the market, and a part may still remain as
pillars in the mine, though not well to be removed
now. In 1872 the yield of the mines was said to be
10,000 tons. The working had been carried on with
long interruptions for half a dozen years at the present
mines ; and before that, probably to a much smaller
extent still, at the now abandoned Hurushiki mines.

At the present mines there are three main gangways
or levels : the Honshiki (or principal one) ; the Shinshiki,
about 72 feet higher ; and the Midzunukishiki (or
drainage level), about 50 feet below the Honshiki. In
May 1873, the Honshiki was nearly 500 feet long ; and
it is now still longer. The Midzunukishiki was begun
nine or ten years ago on the hill side below the Honshiki,
and evidently without the least surveying to find out the
probable position of the bed at that level. It was in
solid rock nearly 60 feet across the beds above the place
where the Honshiki coal would be, if there were no fault
(and none was then supposed to exist). Yet after driving
the tunnel 75 yards in a direction parallel to the strike of
the coal bed, the course was changed to the left, so that the
distance from the bed sought for was much increased. The
tunnel and the mines generally were then abandoned for
some time ; and on their being worked again under the
energetic and comparatively intelligent management of
Mr. Shida, he began at the bend in the tunnel and drove
it straight forward. At length within 50 yards, owing to
a fault (for the course of the tunnel was otherwise about
parallel to the bedding), a coal bed (the Midzunukishiki)
was struck. From that he tunnelled across to the right
and found first a thin bed (the Lower Midzunukishiki)
and at length the Honshiki bed itself ; and now a level is
driven upon that bed there. Mr. Shida also drove an

exploratory tunnel across from near the mouth of the Honshiki level, about 200 feet, and found the three Tateire beds ; but on driving in upon them proved them to be not workable there. He also opened in 1873 the Tariokô drift down in the hollow against the Midzunuki-shiki upon the Kosawa No. 1 (really by a fault the same as the Honshiki bed); but last year, after the level had been driven about 80 feet north-easterly, the bed grew thin, no doubt owing to the nearness of a fault ; and the drift was abandoned. In 1875 a low level drift on the Midzunukishiki coal bed was begun at the Osawa No. 1 exposure; and as the coal proved to be of good quality and about four feet thick it was to be driven forward with energy.

In the summer of 1875 the yield of the mines had be-come extremely small, and there was working room in them for less than 20 men ; for the eastern fault in the two up-per gangways and a local thinning of the bed in the lower gangway had discouraged the driving forward of the levels. It is probable however that such faults here are not very serious ones, and that by tunnelling though the rock to the left (north-westerly) the bed may be found again at no great distance. It is extremely desirable that the three present gangways and a new one beginning at the Osawa No. 2 exposure should be driven forward as rapidly as possible, so as to give, within a year or two, working room for a larger number of miners, and enable the output of the mines to be very much increased at short notice thereafter, a capability that adds greatly to the value of a mining property. A drift should also be begun without delay at the Onkosawa No. 1 exposure and driven easterly, so as to open up as quickly as possible a large amount of water free coal in the high hill there. It would be still better however if the same bed could be found and opened up by drifts half a mile down stream, on either side of the main Chatsunai valley, where the map shows the outcrop to cross.

5. *Shipment.*—As to shipping the coal, navigable streams or canals are quite out of the question ; the only questions are in regard to roads and harbours. The only road near the coal at present is the tramroad (worked by bullocks) from Kayanoma to the present (Honshiki) mines with traces of an old line further up the Tamagawa to the old Hurushiki mines. The tramroad is about two miles (29 cho) long ; and is of three foot gauge and of gentle grade from the sea shore up the Tamagawa valley to within about half a mile of the mines ; then of two foot gauge with three inclined planes up the small branch valleys to the Honshiki. The shipping place at Kayanoma cannot be called a harbor, it is so wholly exposed from north-north-west round by west to south-south-east; and in the present condition of things it is idle to think of shipping there the product of even a single large colliery. It has been proposed to make a long breakwater to give protection against the bad north-westerly winds; and the project was favoured by the only maps that had been made before our survey and that made the sea shore at Shibui to be much further from the mines than it really is. The distance, in fact, by a railroad line around the hill to Shibui would be about the same as by the present tramroad to Kayanoma, and as there is no comparison in other respects between the advantages of the two places for making an artificial harbor, it is needless to discuss the merits of Kayanoma any further.

If the Honshiki bed, the one of the present mines, should be opened at its most convenient neighboring low drainage level, the Osawa No. 2 exposure, not only would a far larger amount of water free coal be opened up than by the present workings, but the necessity for inclined planes in the rail road would be in a great measure, if not wholly done away with. It would be easy to run a tram line round the hill sides with a constant down grade all the way to Shibui harbor, and that too without the need of any tunnel through the hill, which somebody proposed more than three years ago. It is very probable, too, that the Shibui end of the line may be so arranged that the loaded waggons will descend an inclined plane from the hill towards the shore in such a manner as to draw up the empty waggons to a still higher point on the hill, from which by a separate line they could run down hill to the mines again, saving the labor of either horses or locomotives. But in any case the distance would be only about two miles (29 cho). The slightly more distant Furushiki coal could also probably be shipped by a branch of the same tramroad.

The harbor at Shibui and the adjacent one of Chatsunai, although already far better than the Kayanoma roadstead, are too small and shallow for any large shipment of coal. It seems however very feasible indeed to improve them at no great expense, so as to make a very serviceable harbor for shipping coal by steam vessels (of which a great number would not be required) ; although it would be a small harbor. By building a breakwater 375 feet long, in water 25 feet deep at most, from the reefs at the northern end of the Shibui harbor across to an outlying group of large rocks, and connecting or prolonging them by 250 feet more of breakwater, a pretty well protected harbor of about ten acres (12,000 tsubo) of water at least 25 feet deep would be obtained, indeed about six acres of such water would be wholly land locked taking the somewhat distant Motta-no-saki headland into account. The rest of the ten acres would be exposed only through a space of about 30 degrees in a south-south-westerly direction ; and a much larger surface would be protected from the north-westerly storms which are chiefly dreaded·

The materials for building the breakwater are in the high cliffs close to its shore end. There are no streams emptying into the proposed harbor that could have an appreciable effect in silting it up. The entrance would be wide and easy. Indeed, it is an unusually favorable place for making an artificial harbor. The main drawback is that the bottom is of rock and bad for anchors; but in so small a harbor, so well protected, it would perhaps be possible almost always to make the vessels fast to posts on the shore, as is now done with the half dozen large junks that sometimes lie together in the little, but sheltered, Chatsunai harbor.

The outlet of the coal of the Chatsunai valley would naturally be the Chatsunai harbor, at all events; and the distance by a tramroad with a continous down grade through the valley, easy enough to build, would be a little less than two miles from the present fine Onkosawa No. 1 exposure to the sea. It is probable however that the coal could be opened nearly half a mile lower down the valley, and so make the distance less than a mile and a half.

The idea has also been entertained of building a railroad from the coal to Iwanai, which has already some reputation as a place of safety for vessels, although it looks like a very much exposed roadstead. The possibility of building a breakwater there so as to make a much better harbor has been discussed in my report of progress for last season already printed (see pages 23 to 25). In case of building a coal road from Iwanai, it would probably be best to attack the coal near the Onkosawa valley or to the east of it ; and approach it by a short tunnel of probably not more than 1,500 feet in length from the Hattari side of the hill ; or else to sink a deep vertical shaft on the Hattari side, taking advantage of the dip of the coal in that direction. It would probably be much too

difficult to carry the coal by a railroad across the hills from
the Chatsunai valley and still more so from the Tamagawa
valley, and quite impracticable to take a railroad line
round the shore near the sea level.

It has moreover been proposed to make an artificial
harbor at the mouth of the Iwanai or Horikap river.
But the river, though pretty wide, is very shallow, only
six feet at the Horikap ferry ; and so large a stream in a
wide alluvial valley would bring down a great quantity
of silt to fill up any artificial excavations, and cause a
constant yearly expense for dredging. The coal for such
a harbor, too, would best be approached by way of Hattari. But in respect of the artificial harbors no project in
this neighborhood can compare in feasibility with, the
Chatsunai and Shibui one. Its features are shown more
in detail by the hydrographic map of the place.

6. *Maps.*—The map of the coal field (finished and dated
18 May, 1876) is on a scale of $\frac{1}{7000}$, and shows the shape
of the surface of the ground by contour lines ten feet
apart in level, and the shape of some of the more important
portions of the coal beds by similar (but broken and much
more regular) contour lines upon them one hundred
feet apart in level down to 500 feet below water level.
The staked stations of the survey, two hundred feet apart
in straight lines, except a few streams and other crooked
lines, are shown with their distinguishing marks by
which any point may be looked for on the ground. The details of the map are in short much the same as those of our
other surveys. Numerous unexaggerated cross sections
on a scale of $\frac{1}{5000}$ are given to show the dip and structure
at different places ; and several columnar sections on a
scale of $\frac{1}{1000}$ show the different beds in their natural relative position. There is also a small map of Yesso to show
the general position of the Kayanoma Coal Field in comparison with that of the Ishcari coal surveys ; and a little

map of Japan and the surrounding shores to show the position with reference to commerce. The whole covers a space of 3.81 feet by 3.40 feet.

The survey was chiefly made in May and June, 1873, by myself with the aid of Messrs. H. S. Munroe, T. Yamauchi, T. Inagaki, T. Kuwada, S. Misawa, J. Takahashi, T. Kada, I. Bau, T. Saito, Y. Catoö and H. Satoö. In August 1873 Mr. Yamauchi with the aid of Messrs. Kuwada, Takahashi and Saito added to the survey a large portion near Hattari ; and I made the Chatsunai and Shibui hydrographic survey with the aid of Messrs. Yamauchi and Takahashi. In 1875 a small addition was made to the field work by myself with the aid of Messrs. Misawa and Takahashi ; and then the discovery of the faults was made, which had a very important bearing upon the mapping of the geology of the field.

The surveying was done with prismatic compasses and pacing, and the leveling partly with hand levels (especially on the south-eastern half, or towards Hattari) and partly with aneroids. Each surveyor plotted his own lines, and in general drew the contour lines ; and the whole was united into one map by myself. The geological observations were made by Mr. Munroe and myself, and a few also by Mr. Yamauchi. The geological cross sections were drawn chiefly by Messrs. Kada and Misawa and myself ; the columnar sections were reduced by Mr. Kada from and drawings chiefly by myself on a scale of $\frac{1}{100}$. The geological mapping was done under my guidance by Mr. Misawa, who also had charge of making and finishing the final copy of the map, until on account of his illness it came at last into the hands of Messrs. Inagaki and Kuwada.

The hydrographic map is one of a rough survey and was finished and dated in April, 1874. It covers a space of 1,54 feet by 1,93 feet ; and is on a scale of $\frac{1}{1000}$; and includes the whole of Chatsunai and Shibni harbors and the outlying rocks. Without giving contour lines for the very small portion of the land that is shown along the shore, the map gives the shape of the sea bottom by contour lines five feet apart in level. They are based upon numerous soundings made from a small boat rowed slowly and uniformly along a number of lines across from point to point. Mr. Sbida heaved the lead while I watched and recorded the soundings. The outline of the shore is partly taken from my own survey of it, made for the coal map; with the addition of the low water reefs surveyed by Messrs. Yamauchi and Takahashi; and the place of the outlying rocks was determined by my triangulation from the shore. The place of the proposed breakwaters is marked and lines are given to show the outline of the protected water.

I have the honor to be,

Sir,

Your most obedient Servant,

BENJ. SMITH LYMAN,
Chief Geologist and Mining Engineer.

Matsudai, Echigo,
6th September, 1876.

www.ingramcontent.com/pod-product-compliance
Lightning Source LLC
Chambersburg PA
CBHW021942220326
41599CB00013BA/1494